# A practical guide to inspecting

INTERIORS
INSULATION & VENTILATION

## *by Roy Newcomer*

AMERICAN HOME INSPECTORS TRAINING INSTITUTE

# CONTENTS

# INTRODUCTION

My background includes many years in construction and several more as the owner of a Century 21 real estate franchise. In 1989, I started a home inspection company that has grown larger than I ever imagined it could. Training my own staff of inspectors to the highest inspection standards has led to my teaching home inspection seminars across the country and developing study courses, books, and videos for home inspectors. The American Home Inspectors Training Institute was founded as a result of my desire to share this experience and knowledge in home inspection.

The *Home Inspector's Guide* series is intended for both beginning and experienced home inspectors. So if you're studying home inspection for the first time or are using the materials as a refresher, these guides should be of assistance to you.

I've created these guides to include all aspects of home inspection. Not only a broad technical background in home systems, but the other things you need to know in order to perform a *good* inspection of those systems. They lay out technical information, guidelines for the inspection, how-to instructions for inspecting system components, and the defects, deficiencies, and problems you'll be looking for during the inspection. I've also included some advice on how to report your findings to the home inspection customer.

I've been a member of several professional organizations for a number of years, including ASHI® (American Society of Home Inspectors), NAHI™ (National Association of Home Inspectors), and CREIA (California Real Estate Inspection Association). I am a great supporter of those organizations' quest to promote excellence in home inspection.

I do encourage you to follow the standards of the organization to which you might belong, or any state regulation that might take precedent over the standards used here. Use the standards in this book as a general guide for study and apply the standard or state regulation that applies to you.

The inspection guidelines presented in the Inspector Guides are an attempt to meet or exceed standards and regulations as they exist at the revision date of the guides.

There's a lot to learn about home inspection. For beginning inspectors, there are some *hands-on exercises* in this guide that should be done. I'm a great believer in learning by doing, and I hope you'll try them. There are also some of my *personal inspection stories* to let you know what it's really like out there.

The *inspection photos*, located at the back of this guide, are referenced in the text. You'll read the story about each one as you go along. Be sure to watch for my *Don't Ever Miss* lists. I've included them to alert home inspectors to report those defects (if found during the inspection) in the inspection report. If missed, these items are often the cause for lawsuits later. Finally, to help you see how you're doing as you study this guide, I've included some *worksheets.* The answers are given for each one for self checking. Give them a try. Checking yourself can help you lock important information in your mind. There's also a *final exam* that you can complete and send in to us. Many organizations and states have approved this book for continuing education credits. Submit the exam with the required fee if you need these credits.

In total, the *Home Inspector's Guides* cover all aspects of the general home inspection. Each guide covers a major aspect of the inspection, as their titles show:

- *Electrical*
- *Exteriors*
- *Heating and Cooling*
- *Interiors, Insulation, Ventilation*
- *Plumbing*
- *Roofs*
- *Structure*

If you are interested in other titles in the series, please call us at the American Home Inspectors Training Institute to order them. Call toll free at 1-800-441-9411.

*Roy Newcomer*

# INSPECTING INTERIORS, INSULATION, VENTILATION

# Chapter One
# THE INTERIOR INSPECTION

This chapter presents an overview of the inspection of the interior of the home.

## Inspection Guidelines and Overview

These are the standards of practice that govern the inspection of the interior components of the property.

*Guide Note*
*Pages 1 to 4 lay out the content and scope of the interior inspection. It's an overview of the inspection, including what to observe and what specific actions to take during the inspection. Study the guidelines carefully.*

*For convenience, we've included the inspection of fireplaces and wood stoves in the interior inspection.*

| Interior System | |
|---|---|
| OBJECTIVE | To identify major deficiencies in the condition of the interior living structure, including walls, ceilings, floors, windows, and doors. |
| OBSERVATION | Required to inspect and report:<br>• Walls, ceilings, floors<br>• Steps, stairways, balconies, and railings<br>• Counters and a representative number of cabinets<br>• A representative number of doors and windows<br>• Separation walls, ceilings, and doors between a dwelling unit and an attached garage or another dwelling unit<br>• Security bars<br>• Safety glazing in locations subject to human impact<br><br>Not required to observe:<br>• Paint, wallpaper, and other finish treatments on the interior walls, ceilings, and floors<br>• Draperies, blinds or other window treatments, or carpeting<br>• Household appliances<br>• Recreational facilities |
| ACTION | Required to:<br>• Operate a representative number of windows and interior doors.<br>• Report signs of water penetration into the building or signs of abnormal or harmful condensation on building components. |

Most standards of practice provide a good outline of what is to be inspected and what is to be reported during the inspection of the home's interior living spaces. But, of course, not all the details are included. There are many other details that will be presented in this guide.

Although most standards include fireplace and wood stove inspection as part of the heating inspection, we consider them to be part of the interior inspection. The table below reviews the guidelines regarding these items.

| Fireplaces and Wood Stoves | |
|---|---|
| OBSERVATION | Required to identify and report:<br>• Solid fuel heating systems and fireplaces<br>Not required to observe:<br>• Interior of flues, unless readily accessible<br>• Fireplace insert flue connections |
| ACTION | Not required to:<br>• Ignite or extinguish solid fuel fires |

Virtually every space within the living area is inspected, including all rooms, closets, foyers, hallways, and stairwells. It should be noted that the home inspector will be checking **plumbing fixtures and faucets**, the **electrical switches and outlets**, and the presence of **heat and cooling sources** in each room during the interior inspection. These items were presented in great detail in other of our guides. We'll repeat some of that information in these pages as a reminder. Here is an overview of the additional aspects of the interior inspection:

- **Walls, ceilings, and floors:** The home inspector examines the **condition** of walls and ceilings in all rooms and finished spaces, noting **defects** in the plaster, drywall, or other facings and watches for cracking and evidence of leaking. Floors are examined to determine if they're level, show evidence of structural problems, and for deterioration. Particular attention is paid to evidence of leaking and wood rot in the kitchen and bathrooms where plumbing is present. The inspector must keep an eye out for and be sure to report signs of **water penetration** into the building and any signs of abnormal or harmful **condensation** on building components.

The home inspector also inspects **separation walls and**

**INSPECTING INTERIORS**

- Walls, ceilings, and floors

- Interior stairways and balconies

- Counters and cabinets

- Doors and windows

- Fireplaces and wood stoves

*Personal Note*
*"The interior inspection gives you an excellent opportunity to establish a good relationship with your customers. Now you'll be talking about things that everyone understands like the condition of ceilings and windows, take advantage of it."*
*Roy Newcomer*

**ceilings** between a dwelling unit and an attached garage or other dwelling unit for proper construction, wall and ceiling coverings, and safety requirements.

Note that the inspector is **not required to observe paint, wallpaper, other finish treatments on the interior walls, ceilings, and floors, or carpeting**. In general, the home inspector does not comment on decorative touches in a home. These issues are subjective and a matter of taste and play no role in the home inspection.

- **Interior stairways and balconies:** All interior steps, stairways, balconies, and railings are inspected for condition and **safety hazards**. The inspector examines treads and risers for proper measurements, headroom, proper lighting at stairways, and general condition.

- **Counters and cabinets:** The kitchen of the home is thoroughly inspected, including the condition of the countertops and the built-in cabinets. **Cabinets doors** are opened to examine door operation and any hidden defects which may be hiding inside, especially defects such as leaking plumbing and wood rot found underneath the sink. **Drawers** are pulled out and checked for smooth operation.

NOTE: Although most standards of practice don't require the home inspector to observe **household appliances**, we're going to suggest that you do inspect some selected appliances. In fact, we'll outline the procedures that you should follow to test the operation of selected kitchen appliances and exhaust fans in the kitchen and bathroom. These are simple tests to perform and helpful to the customer.

- **Doors and windows:** The inspector operates doors in the home to find defects such as defective hardware, deteriorating finishes, unlevel installation, and water penetration around sliding doors to the exterior. **Separation doors** between the dwelling and the garage are examined for fire resistance and code requirements.

- The inspector identifies the **type of windows** present in the home and opens a **representative number** of them to check operation. The inspector watches for leaking insulated glass, missing hardware, and rotted sills and

---

### NOT REQUIRED TO

- Observe finish treatments on walls, ceilings, and floors

- Observe window treatments

- Observe interior of flues or fireplace insert flue connections

- Ignite or extinguish solid fuel fires

- Observe recreational facilities

---

*Personal Note*

*"Just because the interior inspection seems so much easier than other aspects of the home inspection doesn't mean that you can slack off during it.*

*"First, the interior inspection can provide clues to major problems in the home's structure, in water penetration, and in other systems. You wouldn't want to miss those clues.*

*"Second, there can be many small defects in the interior that you don't want to miss. You'll only have to pay for them later when customers discover your oversights."*

*Roy Newcomer*

sashes. The home inspector is not required to observe or report on the condition or suitability of draperies, blinds, or other window treatments. Window treatments are a matter

- of interior decoration and beyond the scope of the home inspection. Besides, the home inspector's tastes aren't necessarily any better than anyone else's (they could very well be a lot worse).

- **Fireplaces and wood stoves:** Although considered to be part of the heating inspection, this guide includes the inspection of fireplaces and wood stoves. The home inspector will identify the **type** of fireplace and give the fireplace a visual examination to check the condition of its hearth, mantel, firebox, and damper. The inspector is **not required to ignite or extinguish a fire**. When the flue can be seen with a flashlight and inspection mirror, the inspector examines it for creosote and soot buildup. However, this is not always possible, and most standards state that the inspector is **not required to inspect the interior of the flue**.

Some fireplaces have a **metal insert**, which is really a stove with a door on the front. This insert may sit all the way or partially in the fireplace. There can be problems with the connection between such an insert and the chimney, but it's virtually impossible to inspect this connection. Therefore, guidelines state that the home inspector is **not required to observe fireplace insert flue connections**.

Most standards of practice mention one more item. They state that the home inspector does not have to inspect interior **recreational facilities**. This rule is added to cover items such as an indoor swimming pool, spas, saunas, playground equipment and similar items.

## Inspection Tools

The home inspector needs a strong **flashlight** during the interior inspection to light dark areas and to sidelight ceilings and walls when looking for defects. An **inspection mirror** will come in handy when inspecting the interior of fireplaces. For inspecting electrical outlets in each room, use a **GFCI tester** and **neon bulb tester**, which can be purchased at any hardware store.

*Definition*
*A fireplace insert is a metal stove with a door that is totally or partially inserted into a fireplace.*

***Inspection Tools***

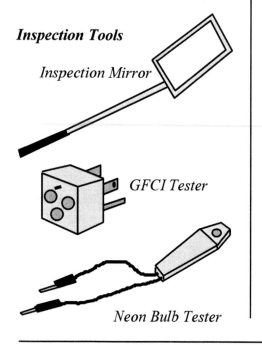

*Inspection Mirror*

*GFCI Tester*

*Neon Bulb Tester*

# Chapter Two
# WALLS AND CEILINGS

**Wall and ceiling facings** are materials such as plaster or drywall that are applied directly to wall studs and ceiling joists. **Wall and ceiling coverings** are materials such as paint and other coatings, plywood, and acoustic tiles that line or finish the facings.

## Plaster Facing

Plaster is a dry powder made largely from **gypsum** (sulfate of calcium) which forms a paste when wet. Other aggregates such as sand or lime can be present in plaster to provide stability, strength, and workability.

Wet plaster is applied in layers, forming a durable wall and ceiling facing. It may be used on solid walls such as brick and concrete block or on wood frame walls. On a frame wall, it was originally applied in 3 layers to a 1" wide by 1/4" thick lath or to a wire mesh nailed to the studs. A scratch coat was applied first, which formed the so-called **key** to anchor it to the lath. A second layer called the brown coat was applied and then a final layer or finish coat. By the 1930's, plaster began to be applied to a **gypsum board lath**, which is a pre-manufactured plaster sheet covered with paper, usually 16" by 48" in size. The gypsum board lath was sometimes reinforced with a **wire lath** at corners and door frames. Then, 1 or 2 layers of plaster were applied.

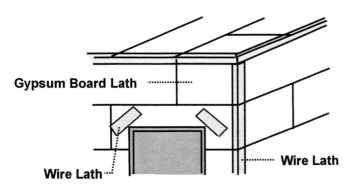

Gypsum Board Lath

Wire Lath

Wire Lath

Plaster is applied with a trowel to the ceiling first, then the walls. The plaster facing does not extend to the floor, and the space left is normally covered with a baseboard. Plaster can be used on arches and other curved or irregular surfaces.

### Guide Note
*Pages 5 to 13 present the study and inspection of interior walls and ceilings.*

*Photo numbers given in the outside margin of the page refer to photos talked about in the text on that page. You'll find the photos in the back of this guide.*

### Definitions
*A _facing_ is the material applied directly to studs and joists to form walls and ceilings.*

*_Plaster_ is a powder made of gypsum and other aggregates that form a paste when wet and a durable surface when applied and dried. _Gypsum_ is a common mineral also called sulfate of calcium.*

*_Lath_ refers to strips of wood, wire mesh, or gypsum board which are attached to studs and joists, forming a base for plaster to adhere to.*

**BULGES**

- Sidelight ceilings and walls to locates sags and bulges.

- You can confirm plaster detachment by tapping and hearing a dull thud.

- You can confirm lath detachment by pushing and sensing movement.

- <u>Be gentle</u>. Don't bring the plaster down.

*For Beginning Inspectors*

*A home inspector needs a good, powerful flashlight for his or her work. It might be a surprise that you'll use a flashlight in the daytime to light ceilings and walls, but you will.*

*Get out your flashlight now and sidelight some ceilings and walls in your own home. Lighting the surfaces from the side will cast a shadow behind any protruding or sagging areas that you'll miss with your naked eye. Try it.*

When inspecting plaster walls and ceilings, the home inspector should watch out for the following defects:

- **Deteriorating plaster:** When plaster is soaked with water, its structure is altered and it becomes **powdery**, causing it to lose its cohesion and strength. Plaster in this condition cannot be re-hardened and must be replaced or covered over with a new facing such as drywall. The home inspector may find powdery areas with no sign of water currently present, but the cause of these areas is definitely water damage.

Deterioration of plaster walls and ceilings can be caused by the condition of the plaster itself. If the original application of the plaster was faulty (from improper additives, extreme temperatures, poor drying, or too much reworking), the finish coat can shrink or lose its adhesion. Extended cold temperatures in an unoccupied house and dampness can cause the lime in the brown coat to disintegrate, resulting in the finish coat cracking, pulling away, and falling off.

Always inspect plaster walls for **water stains**, old or fresh, and investigate the source of those stains. Water may be entering the home through the roof, siding, or windows, be caused by a plumbing leak, or be the result of excessive condensation. Be suspicious of patched areas where repairs have been made.

- **Loose, bulging, or sagging areas:** Use your flashlight when inspecting plaster walls and ceilings, even in the daytime. Get your flashlight up above your head to sidelight ceilings so the light and resulting shadows on the ceiling will alert you to bulges and irregularities in the surface. Stand against the wall and shine your flashlight along the wall for this same reason.

A bulge or sag can indicate that the plaster has become completely **detached from the lath** (lost its key or anchor). This can happen when the plaster deteriorates or when expansion and contraction of the lath forces it loose. If you tap a bulge with the end of a screwdriver and hear a dull thud rather than a crisp sound, it's an indication that the plaster has become detached. *Don't tap too hard on a ceiling because detached plaster is weak and can fall.*

A bulge or sag can also indicate **detachment of the lath itself**. This can be caused by rusted and pulled lath nails. As nails pull out, the weight carried by the remaining nails can become too much to bear. Over time, a large area can become loose and collapse. Lath detachment can be determined by pushing against the bulge to see if it moves. *Don't push at the plaster too vigorously.*

- **Cracks:** Be sure to note any cracks you find in plaster walls and ceilings, even if you can't correctly diagnose their causes. Plaster can show cracks due to some fault in the plaster, normal stress the framing exerts on the plaster, or abnormal movement in the structure of the house. In general, if narrow cracks exist without separation, the cause is not structural. But cracks wider than 1/4" or those with displacement at each side are probably due to structural problems. (More information on structural cracks are presented in another of our guides — *A Home Inspector's Guide to Inspecting Structure*.)

A fine **network of cracks** in the plaster is normally a sign that the original plaster application was faulty, as noted on the previous page, where the finish coat can shrink and crack. There may be cracks in the plaster at the **intersection of different backing materials** such as brick and lath due to their different expansion and contraction rates. Plaster can also crack from the **expansion and contraction of the framework** when there are wide temperature swings or excessive humidity in the home. Plaster over gypsum board may show cracks at the perimeter of the boards in this case. A solution to these types of non-threatening cracks is to cover them with flexible patching materials or wall coverings instead of filling them.

Cracks that signal **structural problems** should be further investigated to try to determine their source. The following are some examples:

— Cracks on interior walls can be caused by insufficient support to the wall and a resulting sag to the wall. Interior walls that are parallel to the supporting girder should be carried on double joists. Those that are perpendicular may need additional support to the joists.

---

INSPECTING PLASTER

- Deterioration and water damage
- Loose, bulging, or sagging areas
- Cracks

---

***Guide Note***
*For information on cracks in the framing and foundation, see another of our guides — A Home Inspector's Guide to Inspecting Structure.*

*Definitions*

*Drywall, which is a pre-manufactured plaster sheet covered with paper, is used as a wall and ceiling facing.*

*Joint cement is a plaster-like paste used to seal the joints between sheets of drywall.*

— Angled wall cracks can be an indication of settlement of the framing or the foundation.

— Ceiling and wall cracks around open stairways can indicate normal stress at these points. More serious cracking with displacement indicates lack of proper structural support in the stairway framing.

— Second-floor ceiling can crack when stressed by a knee wall in the attic. **Truss uplift** is when the bottom chord of a roof truss bows upward during the cold months and returns to its position during the warmer months. Truss uplift can carry the ceiling with it, causing cracking to appear at the junction of the ceiling and walls. A solution to this condition is to conceal the crack with a molding.

— Aging ceiling joists can sag enough to cause the ceiling plaster to crack from the stress. These cracks can run the length of the house along each side of the main girder in the middle of the joist span.

— Tiny cracks from the corners of doors and windows are an indication of a expansion and contraction of the lintel or header.

## Drywall

Drywall, which is basically the same material as plaster, came into use in the 1960's and is still used today. It's also called plasterboard, wallboard, and sheetrock. Drywall comes in large 1/4" to 5/8" thick sheets 4' wide and up to 12' long and is made of a gypsum mixture covered with a treated paper. The edges are slightly recessed so joints can be finished.

Drywall is usually nailed to the framing, although it can be screwed or glued to the wall studs. **Joint cement**, which is a plaster-like paste, and a tape is applied to provide a level, seamless

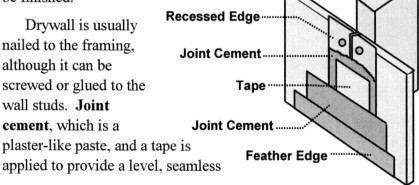

Drywall · · · · · · · · · · · · · · · Wall Stud · · · · ·

Recessed Edge · · · · · · · ·

Joint Cement · · · · · · · · · ·

Tape · · · · · · · ·

Joint Cement · · · · · · · · · ·

Feather Edge · · · · · · · ·

surface. Special techniques are used to finish the edges as shown in the drawing on the bottom of page 8.

Drywall is also used for **separation walls and ceilings** between the garage and the living area, as required by code for fire safety. In some areas, 1/2" drywall is allowed, although other areas require a fire-rated 5/8" drywall to be used.

The home inspector should watch for the following problems with drywall:

- **Deterioration:** Drywall can deteriorate due to water damage just as plaster does. Try to determine the source of the water stains. Sections of deteriorating drywall can be replaced. Tape at the seams of drywall can become loose if exposed to water damage or too much dampness.

- **Nail pops:** These are bulges in the drywall where the nails are backing out of the studs. Nail pops are usually due to the expansion and contraction or shrinkage of the framing. But they can also indicate inadequate bracing in the frame. In extreme cases, the joint cement and tape can pop off and expose the nails. Assuming there are no structural problems, nails can be reset to solve the problem.

If the drywall is too heavy, **dimples** can form around concealed nails. If you push up against the drywall near a dimpled area and you sense a little movement, it can indicate that nailing during the installation of the drywall could have weakened the plaster content in this area. It can also mean that too few nails were used.

- **Loose or sagging areas:** If the drywall on a ceiling is less than 1/2" thick, it can sag. So can damp drywall, which becomes heavier as it picks up moisture, or poorly nailed drywall.

- **Cracks:** Joint cracks in a drywall facing can be the result of expansion and contraction of the framing or from structural movement. The home inspector should distinguish between joint cracks and visible joint seams, which may just be the result of poor workmanship when the drywall was installed. Drywall sheets can crack from structural stresses. It's not uncommon to see the small cracks from the corners of doors and windows from a shrinking header.

---

### INSPECTING DRYWALL

- Deterioration and water damage
- Nail pops
- Loose or sagging areas
- Cracks

---

*For Beginning Inspectors*

*If you have an opportunity to visit construction sites where homes are being built, arrange to stop by during the installation of drywall. Being able to see how joints are finished can be an aid in determining what can go wrong with them later.*

**FACINGS**

• Plaster

• Drywall

• Wood planking and paneling

• Plywood

• Fabricated boards and panels

NOTE: Drywall may have been installed over a plaster wall or ceiling to cover the deteriorating plaster. Drywall must be firmly attached through the plaster to wall studs or ceiling joists, not to the plaster itself. Otherwise, the old plaster can further detach itself and cause the drywall to sag under its weight.

## Other Facings

Other materials are used for wall and ceiling facings as listed here:

- **Wood planks and paneling:** Facings may be tongue and groove planking, which is also called matchboard. This was common in older homes and may still be seen on closet walls and even some ceilings. Today, you might see planking in family rooms. Solid wood paneling may be used in high quality homes.

- **Plywood:** Thicker plywood in veneered finishes or unfinished can be used as a facing fastened directly to the wall studs without a backing. Plywood as thin as 1/16" and 1/8" without a backing would be too soft and wavy to be used as a facing but may be nailed or glued over an existing plaster base. If you find a plywood covered wall, push against it to determine if there is anything behind it.

- **Fabricated boards and panels:** Wood, chips, and fibers can be used to manufacture various types of boards and panels that can be used as facings for walls and ceilings. Masonite is an example. In some cases, the joints between boards or panels are covered with a wooden lath, usually on walls but on ceilings too to complete the pattern.

## A Word about Structure

When inspecting interior walls and ceilings, the home inspector's main concern should *always* be to detect any sign of **structural problems** hiding behind the surfaces. Facings can show stress that indicates settlement or movement of the structure or problems with the framing such as inadequate or damaged supporting members. Investigate further. And always report these conditions or *suspect* conditions if you can't determine the exact cause.

**STRUCTURE**

Pay attention to walls and ceilings for signs of structural problems. Investigate further and report your findings.

## Wall and Ceiling Coverings

As noted earlier, wall and ceiling **coverings** are materials such as paint and other coatings, plywood, and acoustic tiles that line or finish the facings. The home inspector is not required to inspect the finishes on interior walls and ceilings, but it doesn't hurt to be able to point out any particular problems with these finishes and to have helpful information for the customer. The home inspector should pay attention to any defects noted on the surface of walls and ceilings and distinguish between problems with the coverings themselves and other findings that should be reported such as water damage, deteriorating plaster, and so on.

Some wall and ceilings coverings are listed here:

- **Paint:** Plaster, drywall, and plywood should be cleaned and primed before painting (or applying wallpaper). The primer fills pores, seals off constituents in the facing that may harm the paint, and provides a surface for good adhesion.

  Try not to mistake problems in a plaster surface with problems with the paint. **Peeling or flaking paint** usually indicates that no primer was applied. An old plaster ceiling may be peeling because of a hidden layer of **kalsomine**. This was a mixture of glue, water, and pigment that was applied to ceilings and lasted until the glue deteriorated. Kalsomine should have been washed off with warm water and sponges before painting. To repair this condition, the ceiling needs to be scraped back to the powdery kalsomine layer and then washed before repainting.

  Sometimes, chemicals in the plaster can damage the paint, causing a localized or spotty discoloring. Other stains that bleed into new paint usually indicate the facing was not properly cleaned. Using an oil-based paint over wallpaper can dissolve the ink in the wallpaper which bleeds through into the paint. Using a water-based paint over wallpaper can loosen the glue.

  SAFETY CONCERN: Customers may ask about **lead in paint**. The home inspector is not qualified or required to test for the presence of lead, but this information may prove helpful if customers ask. Lead was used for pigmentation and drying agent in oil-based paints until the early 1950's. It was then used only as a drying agent in both flat and

---

> **LEAD PAINT**
>
> There <u>could be</u> lead paint in homes built before 1980. There <u>most likely will be</u> lead paint in homes built before 1950.

*Definition*
  *Kalsomine is a mixture of glue, pigment, and water that was once used as a finish on plaster ceilings.*

**Personal Note**

*"Don't overlook ceilings in closets. One of my inspectors noticed that a drywall ceiling in a downstairs closet had been covered with wood. That's almost certainly a sign that there's been leaking from above. It's a common way for homeowners to cover up the damage to the ceiling.*

*"When the inspector found an upstairs bathroom above the closet, he investigated closely and found a rotted floor beneath the toilet tank. There had been extensive water damage."*

*Roy Newcomer*

gloss oil-based paints until 1978 when it was prohibited entirely in indoor paint. In most cases, lead was never added to latex paints. So, there *may be* leaded paint in homes built before 1980. And there *most likely will be* leaded paint in homes built before 1950.

Lead affects the central nervous system by slowing development and is especially dangerous to children. Lead can be ingested in the form of peeling paint or breathed in as paint dust. Depending on local environmental regulations, lead paint may be encapsulated with fresh paint, polyurethane, vinyl wallpaper, and so on. In some areas, lead paint has to be removed by a professional removal firm since the process itself is dangerous.

- **Wallpaper:** The home inspector should sidelight wallpapered walls to see if the paper is hiding defects in the walls underneath. Uneven areas should be tapped to test for detachment. Keep an eye out for loose wallpaper that can indicate water damage or failing wallpaper glue. Such areas should be pointed out to your customer since repairs would be needed. Point out wallpaper that's been painted over.

- **Stains, varnish, shellac, and wax:** Natural wood planking or paneling is normally finished with a coating that doesn't hide the wood. An indoor wood stain is a colorant designed to bring out the grain or even out the color, but it's not a finish and should be protected by varnish, shellac, or wax. Varnish is oil based and can't be easily removed after cure. Shellac is a natural product dissolved in alcohol and can be washed off with alcohol or diluted ammonia. Wax can be applied to protect a stain or over varnish or shellac. Wax must be removed before a new finish is applied.

- **Plywood:** Thin decorative plywood may be installed over the wall facing as a covering. Push against the plywood to see if it has a proper backing. A plywood covering can be side lighted to spot areas that are loose.

- **Texturing:** Walls and ceilings may have a textured finish achieved by adding sand or other agents to conventional finishes. Texturing of the finish coat of plaster may be done manually with a trowel. **Stippling** is a texturing process where a stipple finish is sprayed over drywall.

- **Acoustic tiles:** Acoustic tiles, popular as a ceiling finish since the 1950's, are typically made of fiber board, but can be made of fiberglass, cork, or mineral particles. They may be plastic coated. Earlier tiles were a foot square. These tiles can be nailed or glued to strapping, installed over a plaster ceiling, or hung in a grid of metal moldings. The acoustic tiles in hung ceilings are larger. Note that installing a hung ceiling lowers the ceiling by 2" to 3".

Photo #1 at the back of this guide shows **water stains on an acoustic tile ceiling**. Note the metal grid holding the tiles in place. It's easy to spot the water stains on this ceiling tile. If you find this situation, be sure to push that tile up to see what's happening behind it. This room was on the first floor, and a second-floor bathroom was right above it. We pushed the tile up and saw evidence of leaking from the plumbing coming from that upstairs bathroom.

← *Photo #1*

## Trim

Most rooms have interior trim including **baseboards** at the intersection of walls and floors and **casings** around windows and doors. Baseboards can be made of wood, plastic, and even tiles or marble. There may be **cornice moldings** at the intersection of walls and ceilings. The home inspector should inspect all the trim and note any missing, loose, cracked, broken, or water stained pieces.

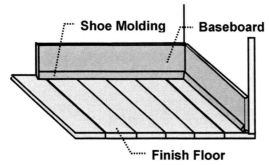

Shoe Molding    Baseboard

Finish Floor

# WORKSHEET

*Test yourself on the following questions.*
*Answers appear on page 16.*

1. According to most standards of practice for the interior inspection, the home inspector is <u>not</u> required to:

   A. Observe a representative number of doors and windows.
   B. Operate a representative number of doors and windows.
   C. Observe draperies, blinds, and other window treatments.
   D. Report signs of water penetration into the building.

2. When inspecting fireplaces, the home inspector is required to light a fire in the fireplace.

   A. True
   B. False

3. What items are <u>not</u> included in the interior inspection?

   A. Closets, foyers, and hallways
   B. Walls, ceilings, and floors
   C. Separation walls between the home and garage
   D. Recreational facilities

4. What causes plaster to become powdery?

   A. Being soaked with water
   B. Detachment of the lath
   C. Settlement of the foundation
   D. Too much reworking upon installation

5. Which type of crack in a plaster facing is an indication of a plaster problem rather than a structural cause?

   A. Angled wall cracks
   B. Network of surface cracks
   C. Cracks around stairways
   D. Cracks above doors and windows

6. How should the home inspector test a bulge in a plaster wall for detachment from the lath?

   A. Side lighting it with a flashlight
   B. Tapping on it with a screwdriver
   C. Banging on it with the flashlight
   D. Pushing on it with the hand

7. Which type of facing would be best suited to curved ceiling surfaces?

   A. Drywall
   B. Plywood
   C. Plaster
   D. Wood planking

8. What is drywall made of?

   A. Compressed paper
   B. Gypsum covered with paper
   C. Kalsomine
   D. Wood chips and fibers

9. What might be the cause of nail pop with a drywall facing?

   A. Expansion and contraction of the framing
   B. Shrinkage of the framing
   C. Inadequate bracing in the framing
   D. All of the above

10. What materials are used for making seamless joints in a drywall facing?

    A. Gypsum and glue
    B. Joint cement and tape
    C. Nails and screws
    D. Shellac and wax

11. Homes built before 1980 <u>most likely</u> have lead in the paint.

    A. True
    B. False

# Chapter Three
# FLOORS

This chapter presents information on the construction and inspection of floors in the home.

*Guide Note*
*Pages 15 to 23 present procedures for inspecting the flooring and floor finishes in the home.*

## Construction

Most homes have a 2-layer floor, consisting of a subfloor and a finish floor above it. The subfloor may be softwood tongue and groove planking, laid either perpendicular or diagonally to the joists. Today, most

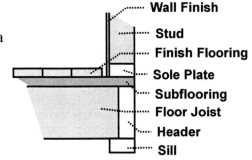

homes use **5/8" thick plywood or particle board sheets** for subflooring. These sheets are laid perpendicular or diagonally to the joists. If the subfloor is laid perpendicular, then the hardwood finish floor is installed parallel to the joists. If the subfloor is on the diagonal, then the finish floor is laid either perpendicular or parallel to the joists.

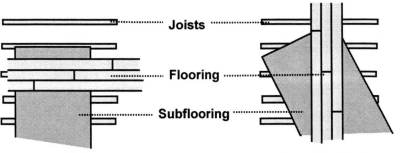

**Subflooring Perpendicular, Finish Flooring Parallel**

**Subflooring Diagonal, Finish Flooring Parallel or Perpendicular**

The meeting edges of the plywood or particle board sheets are nailed to the joists. A structural adhesive may be applied to the joists to add strength to the connection. If the upper layer is unfinished plywood, the rooms are meant to be carpeted or covered with a covering such as vinyl sheets or tiles. The home inspector should inspect 2-layer flooring from above and below where possible. In some cases, hardwood floors are put in without a subfloor. You'll want to check from below first before you assume that a subfloor is present.

In slab-on-grade construction, the subflooring and finish flooring is laid over a poured concrete slab which is generally 4" to 6" thick. Because the slab is not visible, the home inspector must learn to detect slab settlement, cracking, shifting, and water penetration through the slab by the evidence shown in the finish flooring.

## Wood Finish Flooring

Wood finish flooring in a home may be hardwood or softwood, left exposed or covered over.

- **Hardwood flooring** is usually made of oak, although other dense woods are used such as beech, birch, hard pine, maple, pecan, and walnut.

  In a **tongue and groove floor** laid over a subfloor, the hardwood boards are normally 3/8" thick and 1 3/4" wide, but may be thicker and wider in higher quality homes. The boards are laid over the subflooring as shown in the drawings on page 16. In tongue and groove flooring, the top edge of each board is slightly wider than the lower edge. When boards are toe-nailed in place, the top edges fit tight with no nails showing and the lower edges have a small gap.

  Where hardwood flooring is laid without a subfloor, boards should be at least 1/2" thick to 2 1/2" wide to provide the needed strength and support. A hardwood floor may be made of **planks**, which are simply boards more than 3" wide.

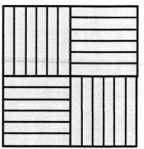

**6" Parquet Squares**

Another style of hardwood flooring is **parquet**. This type of floor consists of 6" squares, each made up of six 1" strips of wood. The squares are laid at right angles to each other. Originally, these strips were cut and laid on the job, but modern parquet squares are pre-assembled and glued to a plywood backing. The squares are then nailed or glued to the subfloor.

- **Softwood floors** are commonly pine, but may be fir or cedar. When softwood is used as a finish flooring, 1" thick by 4" wide tongue and groove boards are used.

Softwood in 1 x 4 or 1 x 6 planks was often used as the **subfloor** under hardwood floors. These planks were spaced with a slight gap and nailed through directly to the joists. Softwood tongue and groove planks were also used as a subfloor under linoleum or other kitchen floor covering. In this case, the top surface was smooth. Sometimes, an underlayment of 1/4" plywood was added between the softwood subfloor and the linoleum or other covering.

The original pine subfloor may have recently been exposed in the kitchen, a practice that's becoming popular. Because the surface is soft and it's hard to keep spills out of the joints, this is not an ideal kitchen floor. Homeowners usually use polyurethane finishes to seal the joints.

**Sanding** of wood floors can get rid of stains and mechanical damage. But it should be noted that there are limitations to the amount of sanding that can be done to hardwood and softwood floors. Although a 3/4" hardwood floor can be sanded several times, a 3/8" one can be sanded only once. A softwood floor should only be sanded once and not to less than 5/8" in thickness when there is no subfloor.

The home inspector should pay attention to the condition of the finish flooring. Water can do extensive damage to wood floors causing boards and parquet squares to **warp, twist, and cup**. Water stains on wood floors may indicate **wood rot**. **Buckling** floor boards are caused by dampness and humidity which cause the wood to swell across the grain and buckle upwards if there is no room to expand sideways. When moisture gets trapped between the flooring and the subflooring, the finish floor can be lifted as each layer expands, warps, and buckles.

A concrete slab may have a hardwood flooring glued over it. Pieces of flooring can pop up if the glue is broken down by moisture in the concrete.

---

**WOOD FLOORS**

- Hardwood tongue in groove and planks
- Hardwood parquet
- Softwood tongue in groove and planks
- Softwood subfloor exposed

---

*Personal Note*
  *"I've seen some exposed softwood subfloors in kitchens where the polyurethane finish needs to be reapplied. The finish doesn't last forever and joints open up again, collecting whatever drops and spills on the floor. I usually warn customers about this so another coat can be applied to prevent water damage."*

  *Roy Newcomer*

## Flooring Tiles

Flooring tiles made of natural materials such as stone, marble, or fired clay can be laid in mortar over a strong subfloor or a concrete slab. Different kinds of tiles include the following:

- **Ceramic or quarry tiles** are hard fired clay tiles that can be glazed or unglazed. They come in variable sizes from 1" to 12" squares and are from 1/4" to 1/2" thick. They're commonly used in kitchens, bathrooms, and entryways where water resistance is important. The most common problems with these floors are cracked tiles due to flexible subflooring and grout deterioration which allows water to seep through the finish floor and damage the wood subfloor underneath.

  The tiles are laid in a mortar base or adhesive over a concrete floor or a stronger than normal subfloor. The tiles are spaced so that grout can be added between the tiles. Current codes require that a 1 1/4" mortar base be used with a galvanized wire mesh laid in it for reinforcement. A water resistant material or sheathing paper should be laid over the subfloor before mortar is added.

- **Slate, stone, and marble tiles** are, of course, natural materials cut to size for use in the home. Their installation is similar to that of ceramic tiles. Because of the weight of the tiles, a concern is that the underlying flooring system is not strong enough to support the tiles without sagging or not rigid enough to eliminate the flex in the floor. Common problems are cracked tiles, deteriorating grout, and staining on marble and other types of stone that can't easily be removed.

- **Terrazzo** tiles are actually a material made up of marble chips set in concrete. The mix is laid in squares bordered by lead beading and then polished for a smooth finish. These types of floors are often found in public buildings such as schools and hospitals, less often in homes.

## Floor Coverings

A variety of other floor coverings are available for installation over a subfloor or even over a finished hardwood floor if the homeowner wishes.

*Definition*

*Terrazzo is a mix of marble chips and concrete that is laid in squares bordered by lead beading. It is polished for a smooth and durable floor covering.*

- **Carpeting** is most often laid over a subfloor and a layer of unfinished plywood, but it can be laid over hardwood floors if the homeowner wishes. The home inspector is not required to report on the condition or quality of the carpeting and customers should be told that. However, you should watch for signs of any problems with the floor underneath and signs of water damage. If carpets have ridges or wrinkles that cause a trip hazard, it can be suggested that the carpet be stretched tight by a carpet professional.

Don't make any assumptions about what's under the carpeting. Even if you find hardwood floors in a bedroom closet, don't assume that the bedroom is hiding hardwood floors under the carpeting. If the customer wants to know, have the owner stipulate in writing what's under the carpeting or offer to lift the carpeting to find out, but don't guess.

- **Resilient floor coverings** include a whole range of tiles and sheet goods that can be laid as a floor finish. Linoleum is a sheet product made of cork and drying oils laid on a cloth backing. Rubber tiles and tiles made from asphalt with inorganic fillers were the first water resistant products made for bathrooms. They were later replaced by plastic floor coverings such as solid vinyl, vinyl-asbestos, and vinyl faced sheets and tiles.

These coverings are laid over the subfloor and an underlayment of 1/4" plywood. Plastic sheets and tiles are laid with adhesives that hold them to the underlayment. Some tiles are available with a peel-and-stick adhesive already applied. Resilient floor coverings may experience the following problems:

— **Cracking and tearing** in sheets can occur from stress when underlayment or subfloor is not tightly fastened and the floor is free to move.

— **Irregularities** in the surface of sheets can occur if the surface underneath isn't smooth. Any gaps or joints in the underlayment can become visible in the covering. When there's water damage in the subfloor and underlayment, warping and delamination can cause the covering to lift. Any ridges or humps in the covering

---

| COVERINGS |
| :--- |
| • Carpeting |
| • Linoleum sheets |
| • Vinyl sheets and tiles |
| • Rubber and asphalt |

will eventually wear out from foot traffic.

— **Snapping and crackling sounds** from the sheet as you walk over it is a sign that the adhesive is defective.

— **Loose sheets and tiles** can occur when moisture penetrates to the adhesive layer, causing the adhesive to lose its bond.

## Inspecting Floors

The first aspect of floor inspection is to determine whether the condition of the floors indicates any **structural problems** in the home. Be sure to inspect floors from above and below, where possible, to investigate the cause of structural problems. The following conditions may be seen:

- **Sloping floors:** An unlevel floor has a continuous slope in one direction. This can be caused by foundation settlement pulling the floor lower at the outer edges of the house. If floors slope inward toward partitions, the condition can be caused when the interior walls shrink more than the outer wood framing and pull the floors down with them. In slab-on-grade construction, where the slab is not attached to or resting on the foundation, a settling slab can pull interior walls down with it.

  Floors should be level. If they're not, it should be noted in your inspection report. Some home inspectors carry a marble or bearing with them to roll on the floor or use a level to confirm a slope. If you find sloping floors, investigate the cause, but don't guess if you're not sure.

- **Uneven floors:** An uneven floor is one that has highs and lows in it, but not a continuous slope. A **hollow** can be caused by the failure of a single joist. When a hollow is present in the floor along an interior wall, it may be that the partition is built between the joists. If the hollow appears on either side of a doorway, it's an indication of poor support for the studs on either side of the opening.

  A **ridge** in an upstairs floor may be caused by a downstairs partition built parallel to a joist. There may be a **bulge** in the floor over a support column, indicating that the column is moving up or the house is moving down. There could be

*For Beginning Inspectors*

*From now on, wherever you go, think about the floors you're walking on. Identify the type of finish or covering, notice its condition, think about the construction underneath, and notice any signs of structural problems.*

a warped or broken joist or an improperly fastened girder under the bulge. Another cause of a bulge can be from an overloaded cantilevered joist, where the joist's interior end is being forced upward.

- **Sagging floors:** This is where there is a low area in the middle of a room. This may be due to poor support in the floor structure. When there are heavy loads such as waterbeds, refrigerators, or pianos, the supporting structures may not be able to hold the load and floors can sag. More support is needed. If an upstairs floor is sagging, check the ceiling below. If it's sagging too, the problem should be looked at by a qualified professional.

- **Deflecting floors:** Floors can have too much upward and downward movement. **Bouncy floors** are usually due to weakness in the joists or a lack of proper bridging. **Soft or springy floors** can indicate a problem between the subfloor and joists — poor support of the subfloor by the joists because of poor nailing or loss of connection. **Spongy** floors can be caused by warped subflooring. Improper spanning are also causes of the above.

Sometimes, concrete slabs have a raised wood floor over them that may be covered with carpeting or resilient tiling. If the floor is spongy or soft when you walk on it, it could be caused by rotted wood in the floor framing where water has seeped in between the slab and the raised floor. It could also be the result of termite damage or it may be that the wood framing is not properly spanned.

- **Noisy floors: Squeaks** in flooring are caused by a poor connection between the subfloor and the joists. Weight on the floor pushes the subfloor down to the joist, and the resulting squeak is caused by nails sliding in and out. You may notice **drumming and rattling** sounds from the floor as you walk across it. These sounds are associated with the joists, not the subfloor. Low frequency sounds are caused by weak flexible floor joists. Higher frequencies are the result of stiffness in the joists.

**Slab settlement and cracking:** In slab-on-grade construction, watch for open joints between the floor and the walls. If the concrete slab settles, it can sink without pulling walls down with it and leave open spaces between

---

**INSPECTING FLOORS FOR STRUCTURAL PROBLEMS**

- Sloping floors
- Uneven floors
- Sagging floors
- Deflecting floors
- Noisy floors
- Open joints at walls

the floor and the interior walls. The walls should be shimmed so that they have the proper support and continue to provide support for structures above them. If the slab settles *and* there is no damage to the foundation, this condition is not a serious structural problem as long as the walls are supported. However, if there are also signs of foundation cracking and settlement, a structural engineer should be called to evaluate the situation.

Inspect the surface of floors over slabs, especially at the edges of the floor, for any indication of **cracks** in the slab. The slab can shrink during cure and pull away from the foundation, leaving cracking along the floor edge. Watch for water damage from moisture coming up through cracks in the slab. Slab cracks may not be easy to notice in the finish flooring. You may be able to feel them underfoot as you walk on the floor, and you may be able to see evidence of them if you sidelight the floor with a flashlight.

The second aspect of the floor inspection is to inspect the **condition of the finish flooring or floor covering**. First, identify the type of floor covering. Inspect from below, if possible, to determine the type of subfloor. Generally, the floor construction as seen from the basement or crawl space is the same throughout the house. Determine if the flooring is put in over a concrete slab. Then watch for the following defects:

- **Water damage and wood rot:** Always look at flooring for water stains or softness that indicate water damage and wood rot to the underlayment and subfloor. *And always report it when you find it.* Be sure to check for wood rot around plumbing in the kitchen and the bathrooms. Don't depend on your eyes alone. Test areas by pushing at them with your foot or get down and press against them with your hands. Another area that should be checked for wood rot is on exterior walls that have an **outside deck or balcony**. When the floor joists extend through the exterior wall to support the deck or balcony, water can seep in. Check along the outer edge of the floor in this area for wood rot.

- **Trip hazards:** Point out trip hazards to customers, including loose floor boards and tiles, torn or raised sheet floor covering, and loose metal or plastic moldings used as

*Personal Note*

*"Homeowners wanting to sell their homes are notoriously clever about covering up defects. One of my inspectors found a new built-in window seat in a room. Everything seemed fine. However, the inspection from underneath revealed a totally rotted subfloor in that area. The owner was trying to hide the problem.*

*"Don't be mislead by new carpeting in bathrooms either. I can't tell you how many times I've seen that trick. A good home inspector will never miss rotted flooring around tubs, showers, and toilets."*

Roy Newcomer

transition pieces between different floor finishes. Trip hazards are easy to notice, repairs are generally easy, and customers appreciate having them brought to their attention. Carpeting can also be a trip hazard if loose and wrinkled.

NOTE: When eyeing the carpeting, be sure to note if carpeting is blocking a heat register or cold air return. With hot water convectors, note if carpeting is blocking air flow.

- **Loose, warped, or buckled wood flooring:** Warping, twisting, cupping, or buckling of floor boards are the result of moisture — either leakage onto the floor or the subfloor or dampness and high humidity — and from being laid too tightly. Try to determine the cause.

- **Loose, torn, or cracked sheet covering:** Inspect linoleum and vinyl sheet goods for any irregularities. Report if you find these defects in resilient sheet coverings. (See pages 19 and 20).

- **Loose, damaged, or missing tiles:** Walk over tile floors, both natural and manmade, feeling for any tiles that are loose. With grouted tile floors, inspect the grout and report if the grout is broken or missing, which would be an indication that water may be getting through the floor.

ABOUT REPORTING: Instructions for using your inspection report to report your findings of floor defects will be discussed later in this guide, beginning on page 62.

---

**INSPECTING
FLOOR COVERINGS**

- Water damage and wood rot

- Trip hazards

- Loose, warped, or buckled wood flooring

- Loose, torn, or cracked sheet covering

- Loose, damaged, or missing tiles

---

# Chapter Four
# WINDOWS AND DOORS

***Guide Note***
*Pages 24 to 33 present information about the inspection of interior windows and doors.*

During the exterior inspection, windows and exterior doors are inspected from the outside. Now, the home inspector is required to inspect interior doors and windows from the inside.

## Window Components

The components showing on the interior of the window are the following:

- **The sash** (upper and lower in the traditional window) is the framework that holds the glass or other material. The top and bottom pieces of the sash are called **rails**; the sides are called **stiles**. The sash is the portion of the window which moves up or down, swings out or in, or remains stationary depending on the type of widows. The sash can be made of wood, aluminum, steel, or plastic.

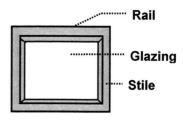

The Sash

- **Glazing** refers to the glass, plastic, acrylic, or polycarbonate within the sash. Each layer of glass or other material is called a **light**. Glazing can be made up of a **single light** (a single pane of glass) or of **multiple lights**, where 2 or 3 thin pieces of glass are sealed together. Glazing is held in the sash by putty or glazing compound.

**Double lights** may be pressed together with no air left between the lights. But in larger windows with double lights, a space between the 2 lights is often filled with an inert gas such as Argon. **Triple lights** may be 3 layers of glass or 2 outer layers of glass with a plastic layer between them. When multiple light windows are cracked or the seal is broken between layers of glass, air and moisture leak into the layers. Leaking multiple lights will discolor and should be reported.

**Muntins** are a grid of cross pieces of wood or lead that hold small panes of glass in a

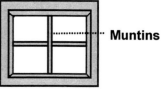

***Definitions***
*A <u>window sash</u> is the framework that holds the glass or other window material called <u>glazing</u>. Each layer of glass is called a light. The <u>window frame</u>, which surrounds and holds the sash, is make up of a top piece called the <u>head</u>, sides pieces or <u>jambs</u>, and a bottom piece or <u>sill</u>. The <u>window casing</u> covers the edge of the frame where it meets the wall.*

multiple pane window. Snap-in muntins made of wood or plastic can be applied over a larger pane to imitate true muntins.

Glazing can be **transparent** (clear) or **translucent** (opaque or cloudy) by design. **Safety glazing** includes tempered glass which shatters into small, smooth edged cubes upon impact. Laminated glass, also with safety glazing, has a sticky plastic inner layer between lights to hold broken pieces of glass when the pane cracks or shatters. Some windows have a wire mesh embedded between the lights for security reasons, but this is not a protection from injury if the glass is broken.

- **The window frame**, which surrounds the sash, holds the sash in place. The frame can also be made of wood, aluminum, steel, or plastic. The top piece of frame is called the **head**, the sides of the

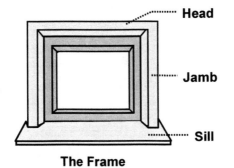
**The Frame**

frame are called the **jambs**, and the bottom piece is called the **sill**. The sill is shaped to be a seat for the sash and to provide weather resistance. The sill can be shallow or deep like the traditional type window sill that you can put a flowerpot on and can extend beyond the casings on each side. Not all style of windows have an extending sill as part of their makeup.

- **The casing** or outermost element of the window covers the edge of the window frame where it meets the wall.

Actually, there may or may not be interior casing around the frame. When casings are present, the casings can vary greatly, sometimes providing decorative touches to the window.

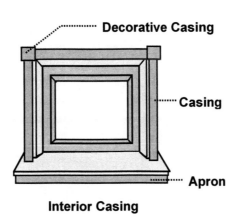
**Interior Casing**

*Definitions*

*The window frame, which surrounds and holds the sash, is made up of a top piece called the head, side pieces called jambs, and a bottom piece or sill. The window casing covers the edge of the frame where it meets the wall.*

---

**Safety glazing** – Report on any glass that is subject to human impact, such as a shower door, if it is not safety glazed

## Types of Windows

The home inspector should identify the type of windows present in the home. Many styles are available, as shown here:

- **Single and double hung windows:** This type of window has 2 sashes. The top sash is on the exterior; the bottom one on the interior. In the single hung window, the bottom sash moves up and down in the frame. The double hung window allows both sashes to move, although you'll often find that the top sash may be painted in place.

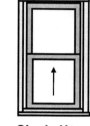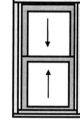

**Single Hung**     **Double Hung**

Early windows were held open in the frame with a counterweight system. A sash cord at each side of the sash goes over a pulley at the top. A counterweight at the end of the cord holds the window in place. Newer types use springs, spiral balances, or coiled steel tape concealed in the side of the sash to serve the same purpose.

Double hung windows may be made of wood, metal, vinyl, or a combination. Glazing may be single, double, or triple lights held in place with putty or glazing compound.

These windows should be opened during the inspection to test the operation of the sash mechanism. The sash cord or other type of mechanism may be broken. The sashes may not move at all or move too freely in their frame. They may not be held in place once opened. When the window is closed, check that the lock holding the sashes in position is operating and holding the sashes tightly to prevent air and moisture infiltration. Newer double hung windows can be pivoted at the bottom so the sash can be rotated out of position in order to clean the outside. This should be demonstrated during the inspection.

Today, double hung windows often come with **storms and screens** permanently mounted in 2- or 3-channel metal or plastic frames. If the system is 2-channel, the sash has to be removed to change the storm and screen panels. In a 3-channel system, also called **self-storing**the screen can be pushed up out of the way. When inspecting these types of double hung windows, be sure to look at the sill. The lower edge of the storm frame should be caulked to the sill, but should have drain holes to allow rain water to run out. Often, the drain holes are blocked and you'll see rotted sills and evidence that excess water has leaked into the house.

- **Casement:** The casement window is hinged at one side and opens either inward or outward, but generally outward. It opens by means of an operating crank usually located at the bottom of the window. Locks or latches on the side opposite the hinges hold the sash tightly when closed. The windows can be made of wood, metal, vinyl, or a

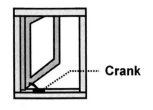

**Casement**

combination and can have single, double, or triple lights. Outward opening casement windows have provisions for a screen or a storm to be installed on the interior of the window frame.

When inspecting casement windows, open the latches and turn the crank to open the window. The crank mechanism can be worn or the gears stripped. Hinges can become corroded. Note if the sash is warped and whether it can close and lock tightly against air and water infiltration. The old metal framed casement windows can rack and make it difficult to close them tightly. Problems should be pointed out to the customer and reported.

- **Awning windows and hoppers:** These are also hinged windows. The awning window is hinged at the top and opens outward. The hopper window, often found in basements, is hinged at the bottom and typically opens inward. Both have a crank operator as a

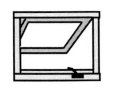 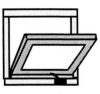

**Awning**　　　　**Hopper**

means of opening the windows. They are in other ways similar to the casement window and have the same types of problems. Inspection procedures are similar as with the casement window.

*Personal Note*
　*"Don't be surprised to find casement windows with cranks that don't work well. One of my inspectors came across one where the gears were entirely stripped. The owner opened the window by pushing it out and closed it by going outside and forcing it closed. Some people can adapt to anything."*
　　　　　　*Roy Newcomer*

- **Sliders:** A slider window is one where the sash moves horizontally in the frame. The slider can have sashes of wood, metal, or vinyl with bearings in the sash or frame for easy operation. Glazing can be single, double, or triple. Sliders should be tested for smooth operation.

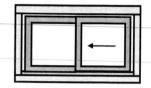

**Slider**

Some sliders are very low quality. They are simply a pane of glass with a handle or knob attached to the surface of the glass and move in a metal or wood track. They are almost never tight and consequently allow air in, making them useless for a cold climate.

- **Fixed-pane windows:** This is a window that doesn't open and close and consists of a large fixed light of glass. The term for this type of window is a picture window. Although some picture windows have a single light, it's more common for them to have multiple lights. The multiple light or insulating windows are commonly called **thermal-panes** or Thermopanes (which is a trade name). You can usually tell if the window has multiple lights by the thickness of the joint between the pane and the frame. A single light is about 3/16" thick, while a multiple light can be from 3/8" to 1" thick. With a double light window, if you put your finger against the glass, there'll be a double reflection of your finger.

Inspect fixed-pane windows for a lost seal between the lights. A leaking thermal-pane window will have condensation between the lights or permanent clouding or discoloration. *Don't ever miss reporting a leaking seal.* Picture windows often have damage on the sill from condensation on the inside surface of the window that flows down onto the sill.

- **Multiple pane windows:** The home inspector may find multiple pane windows with wood or lead muntins. Care should be taken to examine the muntins for breaks, cracks, looseness, and water damage on the horizontal members due to condensation on them.

- **Jalousie:** This window contains narrow strips of glass in a device that allows the strips to move together, lifting outward from the bottom. When inspecting the jalousie window, the inspector should test its operation, paying attention to whether the glass strips will close tightly again. Any damage to the strips themselves should be noted.

**Jalousie**

## Inspecting Windows

During the interior inspection, most standards require the home inspector to do the following:

- Identify the **materials** used in window construction.
- **Look at a representative number** of windows during the inspection.
- **Operate a representative number** of windows during the inspection.
- Report on the **condition** of windows when problems exist.

Watch for and report the following window defects found during the interior inspection:

- **Rotted or damaged sashes, frames, and casings:** Examine each window for any damage to the sashes, frames, and casings. Note especially if any of the framing is out of square or plumb. Windows pulled out of square can be an indication of some movement in the wall. Sashes can come apart at the joints. This often happens when people lift the sash by the top rail rather the pulling it up with the hardware at the bottom. The same is true for sliders where people pull the sash closed rather than pushing. Watch for wood rot on all components. Steel casement windows can rust out and the sashes can rack. Check any **muntins** present for damage and looseness.

- **Rotted or damaged sills:** The sill is probably the most vulnerable part of the framing, and wooden sills often have wood rot. A properly installed window has a sill that tilts toward the exterior for drainage. With self-storing windows, be sure to check the sill carefully. The drain holes should not be caulked over or clogged. Check the wall below the window for water damage to see if water has been leaking in over the sill.

---

**INSPECTING WINDOWS**

- Rotted or damaged sashes, frames, and casings
- Rotted or damaged sills
- Deteriorating paint or finish
- Cracked, broken, or leaking glazing
- Loose, damaged, or missing putty
- Faulty operation
- Missing storm and/or screen

---

*Guide Note*

*The inspection of windows from the exterior is covered in another of our guides — A Home Inspectors Guide to Inspecting Exteriors. Only the inside components are inspected during the interior inspection.*

*Personal Note*

*"One of my inspectors was mystified when he noticed a window sill tilted in toward the room. Then he realized that the window had been installed backwards. Obviously, that was reported."*

Roy Newcomer

- **Deteriorating paint or finish:** Examine all wooden window components to see if a new paint or varnish job is recommended. Windows take a lot of wear and tear and homeowners don't always pay enough attention to deteriorating paint and other finishes. Note if windows that should open have been painted shut.

- **Cracked, broken, or leaking glazing:** Identify the type of glazing present, if possible. Determine whether the glazing is single, double, thermal-pane, and so on. Inspect the glazing for any cracks, breaks, and damage such as BB holes. Don't miss any lost seals in thermal-panes, which can be detected by condensation or discoloration between the lights. A lost seal can't be fixed, so replacement is the only option if the view is seriously impaired.

- **Loose, damaged, or missing putty:** Always check around the window for the condition of putty and glazing compound. Take the time with multiple pane windows to check around each pane for muntins that are loose.

- **Faulty operation:** The home inspector is not required to operate every window, only a representative number in the house. Pick wisely. You may decide to operate one per room or a certain number upstairs and down. However, be sure to operate any window that you suspect won't work properly.

  With double hung windows, keep an eye out for broken sash cords, unworkable locks, broken or missing hardware, windows that are painted shut, that stick, that won't hold their position, and so on. With casement windows and awnings and hoppers, report corroded cranks, missing crank handles, and cranks that don't work.

- **Missing storm and/or screen:** In self-storing storm and screen systems, check that the storm and screen are actually present. The storm could have been broken and never replaced. For casement windows, note the presence of a storm or screen on the interior (depending on the season) and for the portion not installed, ask the owner if it's available. Note its condition if possible.

## Interior Doors

Doors inside the home can be of several types such as flush, louvered, or paneled. They may be hinged, folding, or sliding and made of wood, metal, plastic, or composition.

- **Flush doors** are built of veneers glued to a solid or hollow core. The **hollow core door** is made of a honeycomb or patterned cardboard core which is framed in solid rails along top and bottom and solid jambs along the sides. A piece of wood is located in the knob area for boring out to fit the knob and latch. The construction is covered with a thin outer ply of either painting or finish grade. These doors are light in weight and not very good as sound barriers. Hollow core doors are fairly easily damaged, and it isn't uncommon to see holes in them from overactive kids. The doors can delaminate if the glue doesn't hold.

  Hollow core doors are the most common type used inside homes today and are suitable for interior use except for the fire-rated door required between the living area and the garage (which should be solid core or steel clad).

- **Panel doors** are made of solid components. The rails and stiles (top, bottom, and sides) are grooved in such a way to hold an inset panel. In wood panel doors, the differing shrinking rates can cause the panels to crack along the grain. There are plastic molded panel doors that look like real wood.

- **Louvered doors** have wood, plastic, or cloth slats in a frame. They may be used as room dividers.

- **Pocket doors** are those designed to slide into the walls. Door hardware is concealed in the edge that is exposed at the wall opening.

- **Bifold doors** are designed with hinges in the middle to allow one section of the door to be folded back on the other before the door is swung to the side.

- **Closet doors** can be flush, paneled, or louvered. There may be a single hinged door, a set of hinged pairs, bifold, or doors suspended on tracks so they slide.

In general, interior doors should have the proper **clearances** for a threshold or carpeting and floor coverings. If there isn't a

---

| DOORS |
|---|
| • Flush solid |
| • Flush hollow core |
| • Paneled |
| • Louvered |
| • Pocket |
| • Bifold |

*Guide Note*

*Exterior doors, including entryway doors and garage doors, are inspected during the exterior inspection. This subject is also covered in A Home Inspector's Guide to Inspecting Exteriors. These inspection procedures will not be repeated here.*

return air grill for the warm air furnace in each room, the clearance below doors should be at least 3/4" to allow for air flow.

Interior doors may or may not have **locking capability**. If they do, check that the lock mechanism works. Knob locks generally have a hole in the outer knob so a probe can be inserted to unlock the mechanism. Some knob locks are keyed for additional privacy and security. Keyed knob locks usually have a push-button or other means of disabling the lock function. By the way, try the knobs before you enter a room. You don't want to close a door that locks on you so you can't get out of a room.

## Inspecting Doors

The home inspector should visually examine all interior doors and operate a representative number of them, including opening and closing and checking hardware. Watch for the following conditions when inspecting each door:

- **Poor operation:** Open the doors and check their operation. Doors should hang in their frames with a clearance of about 1/8" on all sides. Something is wrong if they rub, stick, don't close, hang out of square, or are not level. Try to determine what is going on. The door may have swelled from a high moisture content, hinges can be loose, or there may be a structural problem in the framing or wall. Stop moldings are nailed on the inside of the frame to hold hinged doors in position when closed. If the stop moldings are not close enough, the doors can rattle.

  Check to see if a door stop is present behind the door. If not, note if the wall is damaged where the doorknob strikes it.

  Operate closet doors too. If closet doors are hung, check that the small floor guides are in place so the door doesn't swing in and out from its track.

- **Cracked, broken, or damaged components:** Examine the casings, frame, and door itself for condition. Check casings and frame to see if they are out of square, loose, have loose joints, or damaged. Inspect the door *on both sides* for defects such as warping, delaminating plies, holes or dents, cracked panels, scratches, and so on. These deficiencies should be reported.

The home inspector may find a hollow core door that's been cut down to size to fit a smaller frame. Often when this is done, the top or bottom rail may be cut away (since they're not very big to begin with). This is not an acceptable method since the strength of the door is seriously impaired. Instead, a door should be ordered to the size of the frame.

If an interior door has glazing, be sure to inspect the condition of the glazing and report any cracks or breaks. Check the condition of the putty holding the lights in place. With interior French doors with muntins, observe whether any muntins are loose, cracked, or broken.

- **Loose, broken, or missing hardware:** If you haven't checked the hardware on the entryway doors, do so now. Work the locks including the dead bolts to be sure everything is functioning properly. For paired entryway doors with top and bottom slide bolts, check that the bolts extend far enough into up the head and down into the floor.

Test knob locks on interior doors and see if the latch extends into the strike plate. Check that hinges, strike plate, and knob assembly are screwed tight.

ABOUT REPORTING: Instructions for using your inspection report to report your findings of window and door defects will be presented later in this guide. See page 48 for reporting general window findings and beginning on page 62 for reporting room-by-room inspection of windows and doors.

---

### INSPECTING DOORS

- Poor operation
- Cracked, broken, or damaged components
- Loose, broken, or missing hardware

---

*Personal Note*

*"Always look behind doors. One of my inspectors didn't and he missed an electrical subpanel behind a bedroom door."*

*Roy Newcomer*

# WORKSHEET

*Test yourself on the following questions.*
*Answers appear on page 36.*

1. What is typically used for subflooring in new construction today?

    A. 3/8" hardwood tongue and groove boards
    B. 1 x 4 or 1 x 6 softwood planks
    C. Softwood tongue and groove boards
    D. 5/8" plywood or particle board

2. Parquet squares are made of:

    A. Hardwood
    B. Softwood
    C. Terrazzo
    D. Vinyl

3. If you find hardwood floors in a closet off a carpeted bedroom, what assumptions can you make about the bedroom flooring?

    A. The bedroom flooring is plywood.
    B. The bedroom flooring is hardwood.
    C. The bedroom flooring is softwood.
    D. No assumptions can be made.

4. What is indicated if you walk over vinyl sheet flooring and hear snapping and crackling?

    A. Loose nails in the subfloor
    B. Defective adhesive
    C. Open joints in the underlayment
    D. Defective mortar

5. What would <u>not</u> be the cause of a localized bulge in a floor?

    A. Stiffness in the joists
    B. Support column moving up
    C. Warped or broken joist
    D. Overloaded cantilevered joist

6. Hardwood flooring can be glued directly to a concrete slab.

    A. True
    B. False

7. What type of window is shown here?

    A. Hopper
    B. Awning
    C. Casement
    D. Single hung

8. What is a self-storing window?

    A. One that has laminated safety glazing
    B. One that has a permanently mounted screen and storm
    C. One that has multiple panes and muntins
    D. One that has a crank to open and close it

9. Identify the components of a window frame as marked in the drawing.

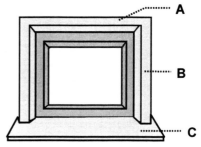

10. The home inspector is required to operate every window and interior door.

    A. True
    B. False

11. The passage door between the living area and the garage should be:

    A. A panel door
    B. A hollow core door
    C. A solid core or steel clad door
    D. A bifold door

# Chapter Five
# STAIRS AND BALCONIES

Included in the interior inspection is the examination of all steps and stairways and balconies within the home.

## Inspecting Stairways

The home inspector should inspect all interior steps and stairways within the home. Codes vary across the country regarding handrail requirements, dimensions of risers and treads, and other aspects of stairways. You would be well advised to check with your own local building codes for information.

When inspecting interior stairways, the home inspector should inspect the following stairway components and dimensions:

- **Stringers:** The stairway is supported on long diagonal supports called stringers which are fastened at both ends to joists and headers in the framework. If the stairway abuts a wall, the stringer can be fastened to the wall studs. A stairway could have 1, 2, or 3 stringers. Stringers are made of wood or metal, but most often wood. Inspect the stringers for cracks, warping, and deterioration of any sort. Watch for any evidence that stringers have shifted or pulled away from the treads.

- **Risers and treads:** Risers are the vertical portion of the stairs. Risers should all be the same height, generally from 7" to 8" high. Some stairways are constructed with **open risers**, that is, with the vertical below each step left open (generally not permitted in newer codes). The treads are the horizontal surface of the stairs. They should be of even depth, generally from 9" to 11" deep including an inch or so for the **nosing** or extension of the tread over the riser.

Inspect risers and treads for uniform height and depth. Unexpected changes between steps can be dangerous and should be reported as a **safety hazard**. Be sure that risers and treads are securely fastened and undamaged. Loose or seriously worn treads presenting a trip hazard should be reported as a **safety hazard**.

NOTE: Interior stairs with open risers can pose a safety

*Guide Note*
*Pages 35 to 38 present the study and inspection of stairs and balconies.*

*For Your Information*
*Check your local building codes for regulations and requirements for interior stairways in the home.*

---

**GENERAL RULE**

Stairs should be uniform. Local codes may allow no more than a 3/8" variation in the width of the treads or the height of the risers. Check your local codes.

---

hazard to families with young children. Youngsters can crawl through the opening between treads and fall through.

- **Handrails and balusters:** A stairway should have a handrail supported by balusters on each open side. With stairs built against a wall, the wall side should also have a handrail, which can be fastened to hangers nailed or screwed to the studs. The handrail should be at least 1 1/2" in diameter and have enough clearance from the wall to permit a good grip. A recommended height for it is from 34" to 38" above the stairs. Balusters should be a maximum of 4" apart to prevent children from falling through. Loose, broken, or missing handrails as well as loose, broken, missing, or too widely spaced balusters should be reported as **safety hazards**.

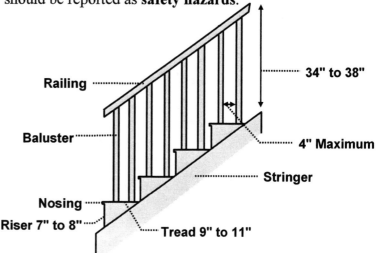

- **Headroom:** Stairs should be at least 34" to 36" wide and allow 6' 8" of headroom above the steps. The home inspector is likely to find that basement and attic stairs, if present, won't meet these dimensions. (Please note that the inspection of the basement staircase is considered to be part of the basement inspection of the home, covered in greater detail in *A Home Inspector's Guide to Inspecting Structure*, another of our guides.)

- **Doors and windows:** If there is a door at the top of the stairs, it should open away from the stairs. If the door opens toward the stairs, there should be a landing present so someone coming up the stairs can't be pushed down when the door opens.

*Worksheet Answers (page 34)*

1. D
2. A
3. D
4. B
5. A
6. A
7. C
8. B
9. A is the head.
B is the jamb.
C is the sill.
10. B
11. C

If a there's a window at a landing or at the bottom of the stairs, it should be at least 36" off the floor. If such a window is lower than that, it should have a guard or grill over it to prevent anyone from falling through.

- **Winders:** When stairways curve or make a turn, pie-shaped treads called winders may be used. Although winders were commonly used in the past, their use is no longer considered as safe as straight stairs. Ideally, there should be a flat landing at a turn. Codes vary as to how narrow the tread of the winder can be. In general, the tread depth should be equal to that of the straight treads at a distance of 18" from the narrow end. The total turn of winders should not exceed 90° with each one turning through no more than 30°. Spiral staircases, of course, are built entirely with winders. But some local codes don't permit spiral staircases as the only means of getting from one floor to another.

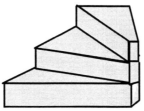

The home inspector doesn't have to report winders as a safety hazard unless their condition is bad. But their presence should be pointed out to the customer.

- **Lack of lighting:** Stairways should be well lighted. Those with more than 4 steps should have 3-way light switches at the bottom and top. (Basement stairs don't require this.)

## Inspecting Balconies

Interior balconies extend from a wall over a living area with no visible connection to the floor below. The balcony can be supported on cantilevered floor joists or on joists or girders spanning 2 bearing walls. Some balconies have supports such as cables, chains, rods, or wood that essentially support the balcony from above or from a back wall.

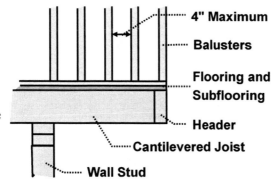

4" Maximum

Balusters

Flooring and Subflooring

Header

Cantilevered Joist

Wall Stud

*For Beginning Inspectors*
*Take notice of any interior stairways and balconies you see. As a mental exercise, try to determine how they're constructed, what members are providing support, and what might cause the structure to become unsafe.*

Balconies should have a handrail and balusters at most 4" apart to protect the safety of persons on the balcony.

Inspect balconies very carefully. The flooring should be undamaged. The whole structure should be securely fastened and sturdy. Walk out to the edge and test the structure for stability. If you notice any shaking, sagging, or tilting, the balcony should be reported as a **safety hazard**. Give the handrail a shake and touch the balusters to see if any of them can be shaken or moved. Any loose or damaged condition that could allow someone to fall through should also be reported as a **safety hazard**.

Pay attention to and report any structural irregularities. When the joists are cantilevered, the downward load at the unsupported end of the joist is reflected by an upward load on the joist at an equal distance from the support point. If the joist is overloaded at its cantilevered end and is pushed downward, there can be a bulge in the floor at the other end of the joist. Or the joist can crush or crack where it's supported by the wall below. Sometimes, the uncantilevered portion of the room can be overloaded, causing the balcony to rise.

ABOUT REPORTING: Instructions for using your inspection report to report your findings of stairway and balcony defects are presented on page 48 of this guide.

# Chapter Six
# FIREPLACES AND WOOD STOVES

Fireplaces today exist for the pleasure of having a comforting fire, certainly not for heat efficiency. Fireplaces pull more heat up the chimney than they provide. Since they use warm house air for combustion, they actually take heat from the house. The use of glass doors on the fireplace and the use of outside air for combustion helps to reduce heat loss. A wood stove, of course, can be an efficient heat source.

*Guide Note*
*Pages 39 to 48 present the study and inspection of fireplaces and wood stoves.*

## The Masonry Fireplace

The conventional masonry fireplace is built with a foundation and footing system as shown in the drawing below. The floor of the fireplace is called the **hearth**, which is a poured concrete form about 4" thick and covered with at least 1" of firebrick, stone, slate, or tile. The hearth is required to extend out in front of the fireplace by 16" and at least 8" on both sides. There's often a hole in the hearth leading to an **ashpit and cleanout** in the basement below so ashes can be removed conveniently.

The **firebox** is the open chamber in which the fire burns. Proper function depends on the relationship between the dimension of the opening and the dimension of the firebox. These measurements can vary according to design. The firebox walls are usually firebrick, but may be made of stone or concrete block with a firebrick liner. Firebricks are mortared with a thin **refractory mortar**. The firebox back wall is often angled toward the front.

At the top of the firebox is the **throat** which is fitted with a metal **damper** that can be closed when the fireplace isn't being used. The throat should be offset from the centerline of the flue, either to the

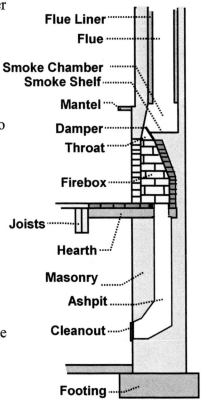

Flue Liner
Flue
Smoke Chamber
Smoke Shelf
Mantel
Damper
Throat
Firebox
Joists
Hearth
Masonry
Ashpit
Cleanout
Footing

*Definitions*
*A <u>hearth</u> is the floor of the fireplace. The <u>firebox</u> is the open chamber in which the fire burns. The top of the firebox is called the <u>throat</u>. A <u>damper</u> is a metal plate that closes the throat when the fireplace is not in use.*
*<u>Firebrick</u> is a special brick designed to withstand high temperatures.*

front or to the back. (The drawing on the previous page shows the throat offset toward the front). The base of the offset is called the **smoke shelf** which provides deflection of downdrafts, snow, and rain. The **smoke chamber** is the area between the damper and the flue. Its sloping wall helps to direct smoke up to the chimney. Some smoke chambers have a smooth parging or stucco finish to encourage the flow of smoke.

The chimney for a conventional fireplace is made of the same material as the fireplace. At one time the chimney was unlined, but since about 1950 a **flue liner** is required. The liner is usually 5/8" thick terra cotta tiles that come in sections from 2' to 3' long. Fireplaces should not share a flue with any other appliance, including another fireplace.

There should not be combustible materials within 6" of the sides of the opening. An example would be a decorative wooden side mantel. Any combustible material above the opening that sticks out more than 1 1/2", such as a wooden mantel shelf, should have a clearance of 12".

Because fireplaces use warm inside air to burn wood, they wastefully draw warm air from the house. One method of making the fireplace more efficient is to bring in **outside combustion air**. Some local codes require it. Outside air is brought in through 4" ductwork from the outside to the floor of the firebox where a damper controls the air flow.

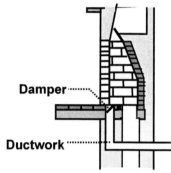

Other devices are used to make fireplaces more efficient. Some fireplaces have **blowers and fans** designed to increase the amount of air blown into the room. Special grates and hollow tubes are other devices used for the same purpose.

NOTE: Some masonry fireplaces have a **metal firebox** of steel plate. The metal firebox is kept 1/2" to 1" away from the masonry and the gap filled with non-combustible insulation. If the gap isn't provided, the metal can buckle and the masonry crack. Don't confuse a metal firebox in a masonry fireplace with zero clearance fireplaces and fireplace inserts mentioned on the following pages.

## Other Types of Fireplaces

The home inspector will come across other types of fireplaces during the course of his or her inspections:

- **Gas fireplaces:** Gas fireplaces were popular around the turn of the century. They're recognizable by their very small fireboxes and decorative borders of marble, cast iron, or ceramic tiles. These old fireplaces cannot be used for wood burning without significant improvements. Today there are natural gas fireplaces on the market, and you may find wood burning fireplaces converted to gas. The new models can't be used to burn wood either.

- **Coal fireplaces:** These were also popular around the turn of the century. You can usually identify an old coal-burning fireplace by the narrow and shallow firebox and the cast iron grate with a pull-out drawer in the bottom to remove the ashes. Some units have heavy slotted covers to put over the opening. In general, it's not a good idea to use a coal fireplace for burning wood, although an examination of the system by a specialist may indicate that everything is in order to burn wood. But recommend the examination in any case.

- **Zero clearance fireplaces:** Since the 1970's, different types of prefabricated or zero clearance metal fireplaces have become available. These are double or triple walled and insulated units that can be wall-mounted or freestanding. A damper is present, but no smoke shelf. Some models have glass doors. The units are usually connected to a metal chimney. They're very light and can be installed without a special foundation. One type of prefabricated unit is called a Heat-o-later, which has inlet grills for cool room air intake and outlet grills above the firebox for heated air release to the room. The wall with the grills can be covered with plaster or drywall.

  Home inspectors have reported many problems with these prefabricated units including corrosion, cracked fireboxes, open flue joints, broken hangers, and displaced metal flues.

- **Fireplace inserts:** An insert is really a stove with a door on the front. This insert may sit all the way or partially in a conventional masonry fireplace and is connected to the

| FIREPLACES |
| --- |
| • Conventional masonry |
| • Gas burning |
| • Coal burning |
| • Zero clearance |
| • Fireplace inserts |
| • Decorative |

fireplace flue. (Inserts are not allowed in zero clearance fireplaces.) Most guidelines state that the inspector does **not have to inspect fireplace insert flue connections**. Customers should be informed of this. There have been serious fire problems due to bad connections with the insert and flue, but it's impossible to see without removing the insert (the insert has to be removed to inspect it properly in any case). Always recommend that a specialist come in to remove and inspect a fireplace insert. Another problem with inserts is that it's difficult to clean the chimney, allowing dangerous levels of soot and creosote buildup.

- **Decorative fireplaces:** From about 1920 to 1940, non-functioning, decorative fireplaces were popular. These fireplaces look real but they're fakes, and there's no chimney associated with the fireplace.. This is why it's so important to *never get sloppy during an inspection and overlook a fireplace*. Imagine reporting that a fireplace is operational only to have your customer discover later that not only doesn't it work but that it's not even a real fireplace.

- **Roughed-in fireplaces:** Sometimes, an opening has been left with a connection to the chimney intended to provide the space for a fireplace to be installed at a later date. Most local codes require the area to be closed off. It's not uncommon for homeowners to try to use them without a firebox and damper system present. Generally, it would cost about $2000 to have the fireplace installed.

## Inspecting Fireplaces

You do not have to ignite a fire in the fireplace during the fireplace inspection, but you are required to inspect the following aspects of the fireplace:

- **The damper** for its operation and condition
- **The firebox lining** for the condition of the firebrick and mortar or other materials
- **The flue** for condition and state of soot and creosote coverage, if possible
- **The facing** around the fireplace for evidence of smoking
- **Clearances** from combustibles

Begin the fireplace inspection by examining the front of the fireplace. Photo #2 shows a **masonry fireplace**. When you start the fireplace inspection, begin by looking over the basics and watching out for the following conditions:

← *Photo #2*

- **Discolored, deteriorating, or loose facing:** Look at the facing just above the firebox. If this area has a dark or black tint or discoloration, it usually indicates that the fireplace releases smoke into the house. A fireplace can smoke for several reasons:

— **Soot and creosote buildup** in the flue.

— **Poor construction** of the fireplace system. For example, the chimney may be too short, the flue may be too small, the firebox could be poorly shaped, there may not be a smoke shelf, and so on.

— A **downdrafting** condition from winds affecting chimney operation. This can usually be corrected by installing a chimney cap on the chimney or a rain cap on the flue.

— A **negative pressure** condition within the home. Some modern homes are sealed so tightly that running a clothes dryer or exhaust fan and using the fireplace can depressurize the home. This condition pulls air down the chimney and causes backsmoking.

Check the facing brick for cracking and mortar condition. Watch for any sign that the facing is loose or moving. The whole facing may move due to a weak floor system that can't support the masonry.

- **Undersized or poorly supported hearth:** The hearth should extend out at least 16" and extend 8" on either side of the opening. If you find an undersized hearth, point it out to the customer with a caution that sparks and coals can fall out. We recommend putting a fireproof rug in front of an undersized hearth. (Note that some prefabricated units won't have a hearth; others have undersized hearths.) Fireplace hearths that sit above floor level should be checked to see if they're properly supported and not tilting or cracking.

- **Improper clearances:** Check that wooden side mantels are at a distance of 6" from the fireplace opening and

*For Beginning Inspectors*
*If you know people with fireplaces, stop in to have a look. Bring a strong flashlight and your inspection mirror and perform an inspection for them.*

mantel shelves at least 12" above the opening. Inspect the condition of wood near the opening. Wood can dry out and char from repeated heatings and has a much lower combustion temperature. Check the mantel shelf by grabbing hold of it and nudging it slightly to test for looseness. But be careful — sometimes you can grab one and it'll be so loose that it falls right off the wall.

The next step in the inspection is to inspect the inside of the fireplace. Inspect the firebox for the following conditions:

- **Cracked, broken, deteriorating, or buckling lining:** First, identify the type of lining that's present such as firebrick or metal. Then use your flashlight to carefully inspect firebrick for cracks, breaks, and any disintegration. Check the mortar for any loose, broken, or missing sections. Watch metal fireboxes for buckling sections and open joints. Open sections in the firebricks or metal could start a fire and should be reported as a **safety hazard** with a recommendation that repairs must be made.

  NOTE: With zero clearance metal fireplaces, examine the unit for any evidence of corroded or cracked fireboxes.

- **Broken, missing, or malfunctioning damper:** Reach in and operate the damper. The damper should open completely and close tightly. Use your flashlight to check if the damper is rusted through, broken, or obstructed in some way. Some older fireplaces may not have a damper, which should be reported with the recommendation that one should be installed.

- **Blower not operating:** If the fireplace has a built-in blower and fan, check it to see if it operates.

- **Dirty flue:** You may or may not be able to see up into the flue, depending on the amount of offset. But try to get your inspection mirror up past the damper and shine your flashlight to see what can be seen. A flue with more than 1/8" of soot and creosote should be cleaned and any obstructions such as a bird nest should be removed. We suggest if you can't examine the flue or very much of it that you recommend that the **flue be evaluated and cleaned** by a specialist. Fireplace flues should be cleaned on a regular basis, depending on use, to eliminate the potential of chimney fires. Some experts suggest cleaning by a professional chimney sweep once for every 1 1/2 cords of hardwood burned.

Most standards state that the home inspector does not have to examine the fireplace flue. That standard was included largely because it's so often impossible to do or do in any significant way. However, we suggest that you make some attempt to look up into the flue, reporting any defective conditions. If the flue is not visible, report that in your inspection report.

NOTE: With zero clearance units, watch for poor connection between components such as open flue joints and displaced metal flues.

Photo #3 in the back of this guide shows a **decorative fireplace**. It looks legitimate, doesn't it? Don't make any assumptions. You must look inside to investigate further and confirm that it's an operating fireplace. Photo #4 shows **what's behind this fireplace**. We were glad we looked!

## Wood Stoves

Inspecting wood stoves involves checking clearances and examining the condition of the stove, smokepipe, and flue requirements. Most standards require home inspectors to inspect a wood stove if it is the **primary heating source**, but leaves the decision to inspect a wood stove to the individual inspectors.

The heavy stove should be firmly mounted on a concrete floor or on a protective pad over a wood floor. The pad should be a non-combustible material and 8" of hollow masonry or a special metal plate with spacers. Stove doors should close tightly, and the stove box should be without cracks or open joints. The wood stove may have a metal or lined masonry chimney and under no circumstances should it share a flue with any other appliance. A wood stove sharing a flue with another appliance should be considered a **safety hazard**. The wood stove produces a great amount of soot and creosote. The flue must be cleaned regularly.

Wood stoves may be listed by the Underwriters Laboratories (UL) and are said to be **UL rated**. Installation clearances are set out in these listings and are typically as those shown here.

---

> **INSPECTING FIREPLACES**
>
> - Discolored, loose, or deteriorating facing
> - Undersized or poorly supported hearth
> - Improper clearances
> - Deteriorating lining and mortar
> - Broken, missing, or malfunctioning damper
> - Blower not operating
> - Dirty flue, open joints, and displaced flues

↖ *Photos #3 and #4*

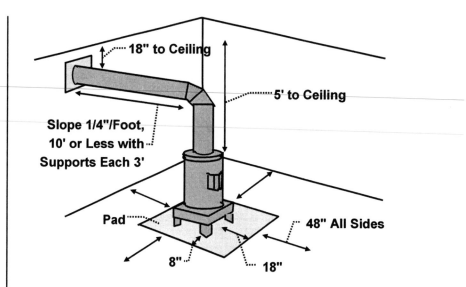

18" to Ceiling

5' to Ceiling

Slope 1/4"/Foot, 10' or Less with Supports Each 3'

Pad

48" All Sides

8"

18"

*Photo #5* ➔

CAUTION: Don't report a wood stove as being safe if you don't know. Sometimes, metal chimneys pass through walls and ceilings and you can't see them to be able to determine if the proper clearances are met. Suggest that a chimney or wood stove specialist be called in to evaluate the safety of the flue. Photo #5 at the back of the guide shows a **metal chimney** passing through a closet. This chimney meets its required 1" clearance from combustibles, but the homeowner has put clothes too close for safety.

## Reporting Your Findings

We're going to catch up on how to report some of the interior items that have been presented so far in this guide, including windows, stairways, balconies, and fireplaces.

The four components of a home inspection are **examination**, **analysis**, **communication**, and a **written report**. Remember that you should continue to communicate with your customer even if you're inspecting "simple" items like windows. Customers still need to hear what's going on. And it may just be that what you consider to be an easy and knowable subject isn't so familiar to the customer after all. Don't preguess what the customer does and doesn't know about. Keep a steady flow of conversation going. Often, it's the small tips that customers will remember best. As you perform the interior inspection, talk to your customer, explaining:

**KEEP TALKING**

Don't go silent during the interior inspection just because the items being inspected are everyday things. Customers still want to know what's going on.

- **What you're inspecting** — the ceiling, walls, floors, the window sill, door hardware, the fireplace damper, etc.

- **What you're looking for** — water damage, sagging plaster, rotted wood, cracked firebrick, uneven stair treads, and so on.

- **What you're doing** — sidelighting the ceiling to find bulges, listening to floor noises, testing the balcony for stability, and so on.

- **What you're finding** — nail pops in the drywall, weak floor joists, a leaking thermal-pane, a loose stairway handrail, a dirty fireplace flue, and so on.

- **Suggestions about dealing with the findings** — replacing a section of drywall, revarnishing the window framing, having a specialist remove and check a fireplace insert, and so on. But with this caution — don't make uneducated guesses about how repairs should be made.

## Filling in Your Report

Every home inspector needs an inspection report. A **written report** is the work product of the home inspection, and every home inspector is expected to deliver one to the customer after the inspection. Inspection reports vary a great deal in the industry with each home inspection company developing its own version. Some are considered to be excellent, while others are not very good at all. A workable and easy to use inspection report is important for a home inspector in terms of being able to fill it in. Of greater importance is its thoroughness, accuracy, and helpfulness to the customer. We can't tell you what type of report to use, but let's hope it's a professional one.

The **Don't Ever Miss** list presented on the next page is a reminder of those specific findings you should be sure to include in your inspection report. We list these items after years of experience performing home inspections. Missing them can result in complaint calls and lawsuits later.

When you're filling in your inspection report, be sure to put in enough detail so your customer knows what your findings were, even if the report is read at a later date.

*Report Available*

*The American Home Inspectors Training Institute provides both manual and computerized reports. These reports include an inspection agreement, complete reporting pages, and helpful customer information.*

*If you're interested in purchasing the Home Inspection Report, please call us at 1-800-441-9411.*

- Windows: rotted framing, broken or leaking glass, loose putty, not operating

- Stairs: Damaged stringer, treads, and risers, loose or missing handrails, balusters too far apart, windows unprotected

- Balconies: Loose or missing handrails, balusters too far apart, instability.

- Fireplaces: Smoky condition, improper clearances, deteriorating lining and mortar, broken or missing damper, and dirty flue

- Wood stoves: Improper clearances, sharing flue with another appliance

Here is an overview of how to report on selected items covered so far in this guide (reporting other items is described on pages 62 and 63):

- **General window overview:** You should have an area in your inspection report for recording general window information. That's the place to record the window materials (wood, vinyl, metal) and types of windows (casement, double hung, etc.) you've found in the house. Be sure to record any evidence of leaking of *any* insulated glass window in the home. Make a note of cracked glass and missing putty too.

- **Stairs and balconies:** State the condition of the stairs and interior balconies, using a satisfactory, marginal, poor designation if you wish. Note defects that you've found and don't miss writing about safety hazards such as missing railings, balusters too far apart, and unstable balconies.

- **Fireplaces:** First, identify the type of fireplace (masonry, wood burning, etc.) If there's more than one, identify each and give its location. If you've tested the built-in blower, don't forget to mention that and indicate whether it's operating. We suggest that if you haven't seen much of the flue or none at all that you always write *Recommend having flue cleaned and evaluated* just to be on the safe side. Note other defects you've found. Pay special attention to any safety hazards you've found. And be sure to indicate if the fireplace is roughed-in only and not a real fireplace.

- **Wood stoves:** If you've found a wood stove that serves as a primary heating source and you've inspected it, be sure to record your findings in your inspection report.

- **Safety hazards:** Never overlook reporting safety hazards. It's a good idea to highlight them by reporting them on the page of the inspection report dealing with the item, and then listing them again on a summary page of your report.

— Uneven or loose risers and treads on stairs
— Loose or missing handrails, balusters too far apart, or instability of stairs and balconies
— Cracked lining in fireplaces and wood stoves, improper clearances, open joints in a metal firebox, sharing flues with furnace, and so on.

# WORKSHEET

*Test yourself on the following questions.*
*Answers appear on page 50.*

1. What is an acceptable distance between the tread of a stairway and its handrail?

   A. 9" to 11"
   B. 25" to 30"
   C. 34" to 38"
   D. 6' 8"

2. Which stairway structure may <u>not</u> be allowed by local building codes?

   A. Open risers
   B. Stairways abutting the wall
   C. Balusters spaced at 4" apart
   D. Winders

3. Which of the following conditions would <u>not</u> represent a safety hazard on an interior balcony?

   A. A loose handrail
   B. A tilt to the balcony
   C. A balcony that shakes when you walk on it
   D. Cantilevered floor joist construction

4. Which of the following conditions might cause a fireplace to smoke?

   A. An undersized hearth
   B. A missing damper
   C. Soot and creosote buildup in the flue
   D. Bringing in outside combustion air

5. What is a fireplace insert?

   A. A prefabricated metal fireplace that's wall-mounted or freestanding
   B. Ductwork bringing in outside combustion air to the fireplace
   C. A fireplace blower and fan unit
   D. A stove that sits in or partially in the fireplace

6. Identify the components of a masonry fireplace as marked in the drawing.

7. When should the home inspector recommend a specialist be called in to clean and evaluate a fireplace flue?

   A. Only for a masonry fireplace
   B. When the flue is not visible or hardly visible
   C. For every fireplace inspected
   D. For none of the fireplaces inspected

8. What fireplace condition would <u>not</u> be considered a safety hazard?

   A. An open joint in a metal firebox
   B. Burning wood in a roughed-in fireplace
   C. Open sections in the firebrick
   D. A broken damper

9. Which statement is true?

   A. A fireplace can share a flue with another fireplace.
   B. A fireplace can share a flue with a wood stove.
   C. A fireplace should have its own flue.
   D. A wood stove can share a flue with a gas furnace.

10. What does *UL rated* mean?

    A. A wood stove is listed with the Underwriters Laboratories.
    B. A wood stove is correctly installed.

# Chapter Seven
# ROOM-BY-ROOM INSPECTION

**Guide Note**

*Pages 50 to 63 present an overview of the interior inspection of the house, room by room.*

The interior inspection includes the careful inspection of each room in the home. The table below gives an overview of what must be inspected in each particular room. Some items such as the heating, electrical, and plumbing aspects relating to these rooms will be reviewed in these pages only briefly.

| Room-by-Room Inspection Overview | | |
| --- | --- | --- |
| **Kitchen** | **Bathroom** | **Other rooms** |
| • Ceiling, walls, floor | • Ceiling, walls, floor | • Ceiling, walls, floor |
| • Doors and windows | • Doors and windows | • Doors and windows |
| • Heat source | • Shower and tub surround | • Heat source |
| • Electrical: fixtures, switches, outlets | • Heat source | • Electrical: fixtures, switches, outlets |
| • Countertops and cabinets | • Electrical: fixtures, switches, outlets | |
| • Plumbing: sink, faucets, pipes, water flow and drainage | • Plumbing: sink, tub, shower, toilet, faucets, pipes, water flow and drainage | |
| • Appliances and fan | • Fan | |

It's important to develop a **routine** for conducting the inspection in each room so that nothing is overlooked. The routine may be inspecting certain items in order each time — for example, first the ceiling, walls, and the floor, then the windows and doors, then electrical, and so on. Another approach is to start by inspecting the door and then proceeding around the room clockwise or counterclockwise examining walls, windows, heat registers, outlets, and so on as you pass by. The important thing is to do it the same way every time and not miss anything.

## Electrical Review

During the interior inspection, the home inspector will examine the **electrical components** in each room. The inspector should determine if there is power and lighting available to each area, if there are enough outlets and they're properly wired and grounded, and if GFCI's are present to protect people in those

**Worksheet Answers** *(page 49)*

1. *C*
2. *A*
3. *D*
4. *C*
5. *D*
6. *A is the damper.*
   *B is the firebox.*
   *C is the hearth.*
   *D is the ashpit.*
   *E is the smoke chamber.*
7. *B*
8. *D*
9. *C*
10. *A*

areas near water. The home inspector is required to:

- Operate a representative number of installed **lighting fixtures and switches** throughout the home.
- Check for the **presence of an outlet** in each room and long hallway.
- Check at least one outlet per room for **current**.
- Check for **polarity and grounding** of all outlets within 6' of water.
- Check for the **presence of GFCI's** where required and **operate and test all GFCI's**.
- Look for electrical **defects and violations**.

To sum it all up, when you inspect a room, you should first turn on installed lighting fixtures. Then check for an outlet and test it for current. In the kitchen and bathroom, check the outlets near water for polarity and grounding and for the presence of GFCI's. Then test the GFCI outlets for operation *twice* with the following methods:

1. Insert the **GFCI tester** into the outlet. If the circuit is correctly wired, the 2 top lights on the tester will be on. Then **push in the black button** at the top of the tester. This should trip the GFCI and interrupt the circuit, and all lights on the tester will be off. If the 2 top lights remain on, either the GFCI is not working or the circuit is incorrectly wired. If the GFCI worked, push in the reset button on the GFCI outlet.

**Lights Off
GFCI Functioning**

**Lights On
Not Functioning**

2. Next, remove the GFCI tester and push the **test button on the outlet** to test it again. If the reset button pops out, the GFCI is working. Push the reset button on the GFCI outlet to reset the circuit.

CAUTION: You may find large appliances such as refrigerators, freezers, or washers plugged into GFCI circuits when they shouldn't be. Freezers and refrigerators in the garage plugged into GFCI outlets are a special problem. After the home inspector trips that GFCI, the freezer will have to restart and the motor surge can trip the GFCI again after the inspection. That

**ELECTRICAL FOR
EACH ROOM**

- Turn on installed lights.
- Check for an outlet.
- Test one outlet for current.
- Test outlets near water for polarity and grounding.
- Test GFCI's.
- Look for defects and violations.

*Guide Note*
  *A Home Inspector's Guide to Inspecting Electrical* presents complete information on the electrical inspection. The information presented here includes only a small portion relating to room by room inspection.

GFCIs wired to 2 wires (open grounds) are legal, but will not trip with the tester.

*Definition*
  *A GFCI is a ground fault circuit interrupter that trips after a ground fault is detected, stopping the flow of electricity in a circuit.*

*"I've tested GFCI's in
several garages that had freezers
plugged into them. Very often
the GFCI will trip again after
the inspection and I'll get a call
about spoiled food. And let me
tell you, the freezers involved in
my inspections always seem to
be holding some very exotic and
expensive food like salmon from
Norway. Paying to replace food
like that can get expensive.*

*"Now I always leave the
owner a note cautioning them to
watch the freezer because the
GFCI may trip again. I also tell
the owner the freezer shouldn't
be plugged into a GFCI outlet.
It saves me money."*

*Roy Newcomer*

**Definitions**

*Reverse polarity is where
the hot wire is wired to the large
slot in an electrical outlet and
the neutral wire is wired to the
small slot, the opposite of how it
should be done.*

*An open ground means that
the outlet is not grounded. A 2-
slot outlet may not be grounded,
while the 3-slot outlet should be
grounded.*

means spoiled food and a complaint to the home inspector. If
you come across this situation, leave a note for the owners
informing them to check that the GFCI hasn't tripped again after
the inspection.

Any **2-slot outlets** will have to be tested with the **neon bulb
tester**. This is a small tool with a light attached between 2 long
prongs (shown in the drawing on the next page) that can be
inserted in the outlet slots and against the outlet screw to get
readings. The table below shows how to test 2-slot outlets with
the neon bulb tester.

| 2-Slot Outlets | |
|---|---|
| **Neon Bulb Tester**<br>Off = ●    On = ○ | **Condition** |
| | **Correct wiring** where hot wire is wired to small slot and neutral wire is wired to large slot. |
| | **Reverse polarity** where hot wire is wired to large slot and neutral wire is wired to small slot. |
| | **Inoperative** for any number of reasons but basically not providing power to the outlet. |
| | **Open ground** where checking small and large slots shows that outlet has power, but checking each slot against screw shows open ground. |

Test **3-slot outlets** with both the **GFCI tester and the neon
bulb tester**. This is a double check of polarity, open ground, and
inoperative outlets.

Be sure to test all **outlets near water** for polarity and open
grounding. Outlets near water with these problems can be
especially dangerous and should be reported as a **safety hazard**
if found. Check the outlet near the main panel for grounding.

Sometimes this outlet isn't grounded.

| 3-Slot Outlets | | |
| --- | --- | --- |
| **GFCI Tester**<br>Off = ● On = ○ | **Neon Bulb Tester**<br>Off = ▬ On = ⬭ | **Condition** |
| (GFCI: top-left ○, top-right ○, lower ●) | (neon bulb tester diagram) | **Correct wiring** where hot wire is wired to small slot and neutral is wired to large slot. |
| (GFCI: top-left ○, top-right ●, lower ○) | (neon bulb tester diagram) | **Reverse polarity** where hot wire is wired to large slot and neutral to small slot. |
| (GFCI: top-left ○, top-right ●, lower ●) | (neon bulb tester diagram) | **Open ground** where the outlet is not grounded and no current flows between the upper slots and the grounding slot. |
| (GFCI: top-left ●, top-right ●, lower ●) | (neon bulb tester diagram) | **Inoperative** for any number of reasons, but basically not providing current or power to the outlet. |
| (GFCI: top-left ○, top-right ○, lower ○) | (neon bulb tester diagram) | Many conditions but usually means there is a **weak ground** which indicates a leak of current from the outlet. |

Don't ever miss noting the absence of power to any area or checking at least one outlet per room, including other rooms such as the utility room.

## Plumbing and Heating Review

During the interior inspection, the home inspector will also examine the **plumbing components**. In general, the inspector will determine if plumbing fixtures and faucets are in good condition, if functional water flow and drainage are acceptable, and if there are any leaks in the plumbing system. The home

**TESTING OUTLETS**
- Correct wiring
- Reverse polarity
- Open ground
- Inoperative

***For Beginning Inspectors***

*This is a good time to buy yourself a GFCI tester and neon bulb tester and get some practice.*

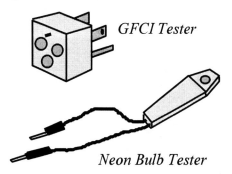

*GFCI Tester*

*Neon Bulb Tester*

*Use these tools to test all your own outlets first. Then carry the testers with you for several days and test outlets at other people's homes. Always be aware that you may be turning off power to important outlets downstream of GFCI's. Ask first.*

## PLUMBING IN KITCHEN AND BATHROOMS

- Turn on all faucets.

- Check for hot water.

- Check for functional flow and drainage.

- Check visible piping.

- Flush each toilet.

- Examine all fixtures for condition.

- Look for leaks.

## HEAT IN EACH ROOM

- Turn up the thermostat.

- Check for heat source.

- Examine condition of heat outlets.

- Confirm heat.

inspector is required to:

- **Operate all plumbing fixtures** in the home, turning on each faucet in sinks, tubs, and showers and flushing each toilet.

- Check for the **presence of hot water** at hot water faucets.
- Check faucets for condition and **functional flow**.
- Inspect all fixtures for condition and **functional drainage**.
- Check visible **piping and traps** under sinks for proper venting.
- Look for any evidence of **leaks** in the plumbing, old or new.

In summary, the home inspector runs the water at the kitchen sink, checks water pressure and drainage, checks the faucet and sink condition, and inspects the trap underneath. In the bathroom, each fixture including the toilet is inspected for condition. Water pressure can be tested by turning on sink and tub faucets and then flushing the toilet. The inspector checks sink, shower, and tub for condition and drainage. The toilet is inspected for operation.

The home inspector will also inspect certain **heating components** during the interior inspection. The inspector should manipulate thermostats to turn the heating system on while in the living area and then observe heating in each room. Registers, radiators, and convectors are inspected and the presence of heat confirmed. The inspector is required to:

- Turn up the **thermostat** to start the heating system.
- Observe the presence of the **heat source** in each room.
- Inspect the **condition and operation** of registers and returns, radiators, and convectors, confirming heat.

Please note that complete information on the inspection of the plumbing and heating systems is presented in other guides in this series. The information presented in this guide on the interior inspection is only a small portion relating to room by room inspection.

## Inspecting the Kitchen

Here's a routine that may be followed during the kitchen inspection:

- **Installed kitchen lighting:** It's a good idea to start the inspection of any room by trying the light switch by the door. Get into the habit of flicking the switch upon entry.

- **Kitchen perimeter and ceiling:** Begin your inspection by circling the perimeter of the kitchen and examining the walls, windows, and doors as you go. Decide which windows you'll check for operation. Pick wisely. Sometimes, the window at the sink is forgotten, and it may be neglected because of its position. Use your flashlight to sidelight walls and take the time to sidelight the ceiling from several locations in the room to locate any bulges or deterioration of the facing. When checking each wall, notice the presence of a **heat source** (register, radiator, or convector) and cool air return in the kitchen. Notice the presence of **outlets** other than countertop outlets. Test one of the wall outlets for current.

- **Floors:** As you made your circle around the room, you should have kept any eye open for any problems with the floor at the junction with the walls, at doorways, beneath windows, and so on. Now walk back and forth across the room and check it carefully for structural problems, squeaks and soft areas, deterioration or looseness in the finish flooring or covering, and any evidence of water damage or wood rot. Check for evidence of rotted subflooring where water damage is likely to occur such as in front of the sink and dishwasher. Keep the floor in mind as you look under the sink and inside cabinets in the next step.

- **Kitchen cabinets:** Next, turn your attention to the kitchen cabinets. Photo #6 in this guide shows **common kitchen cabinets** found during an inspection. Be nosy. Open all cabinet drawers, checking to see if they slide in and out easily. Drawer handles and fronts can be loose. Missing, worn, or broken slides may let the drawer drop when it's partially open.

  Open all cabinet doors, which should open easily, close tightly, and have secured hinges. Notice if interior shelves and partitions are in good shape or damaged or missing.

---

**AROUND THE KITCHEN**

Develop a routine for inspecting the kitchen and follow it religiously. Set it up so you don't forget any of the inspection items on your list.

---

← *Photo #6*

*Photo #7* →

Photo #7 shows **electrical work** passing through a kitchen cabinet. This is not an uncommon practice in kitchen wiring. If you find a similar situation, determine if the wiring poses any safety hazard. With upper wall-hung cabinets, see if any are pulling away from the wall. Fastenings should be secure. You can push up on them from below to check. With lower cabinets, remember to check the condition of the flooring inside the cabinet.

- **Countertops and outlets:** Whatever material is used for the countertops, it should be unbroken, tightly fastened, and not have surface damage such as scratches, dents, burn marks, or holes. Some homeowners will arrange items on the countertop to cover up damage, so it's a good idea to move things around to check to be sure. Lean on the edge of the counter to see if the counter is secure to the base cabinet. You can discover any looseness of laminates by pressing down on the surface. Look around the sink for any evidence of wood rot. With tile countertop finishes, examine them for any damaged or missing grout. The home inspector should be able to give an overall rating of satisfactory, marginal, or poor as to countertop condition. It's not necessary to write up every scratch in your inspection report, but you should mention them in your conversation with the customer.

*For Beginning Inspectors*
*Get out your flashlight, GFCI tester, and neon bulb tester. Perform a complete inspection of your own kitchen as outlined here. Repeat the inspection in a different order until you find one that makes sense to you.*

*Photo #8* →

As you're checking out the countertops, pay attention to the outlets at that level. Kitchen countertop outlets within 6' of water should be grounded, so the home inspector should test them. Photo #8 shows a **GFCI outlet** near the sink. Check for the presence of a GFCI near the sink and test those that are present. If the outlet near the sink doesn't have a GFCI, recommend that one be installed. Also, if the home has under-the-counter lighting, turn it on during this part of the inspection.

*Photo #9* →

- **The sink:** Check the sink for general condition, including cracks and other damage. Turn the water on at the sink and test first for hot water. Watch the faucet for leaking and examine its condition. Photo #9 shows a **kitchen sink**. Let the cold water run like this while you check for functional flow and drainage. This photo also shows a

**sprayer** that should be checked. Often, these sprayers leak or are clogged with mineral deposits and don't work properly. If there's a **garbage disposal** present, turn it on just long enough while the water is running to confirm its operation.

While the water continues to run, look under the sink for any signs of leaking. Photo #10 shows **piping** under the sink, which should be checked for condition. Metal piping should be examined for rusting and corrosion. For proper venting, a P-trap should be present as shown in the photo. Photo #11 shows an illegal **S-trap** that is not acceptable and should be reported. Since the S-trap has no vent, it can let sewer gas into the home. Inform customers that water should be run into the S-trap to be sure the water seal is intact. Ideally, the S-trap should be replaced. Many older homes will have S-traps, which will need replacing when remodeling takes place. By the way, be sure to inspect the flooring under the sink for any signs of wood rot.

- **Appliances:** Assuming that the large kitchen appliances are being sold with the home, we suggest that you test the operation of the dishwasher, the kitchen stove, the refrigerator, and any exhaust fans present.

— **Dishwasher:** Photo #12 at the back of the guide shows a **common built-in dishwasher**. It's always a good idea to ask the homeowner first if the dishwasher works and is hooked up. *Always look inside before you test the dishwasher.* You never know what may be stored in there (sometimes stuffed teddy bears). You won't have to run the dishwasher through a full cycle. A short cycle should be enough to tell if the sprayers work and the motor runs. Check to be sure it doesn't leak on the floor or under the sink. If the discharge hose is connected to the sink drain, it should be water tight and low enough so dishwasher water doesn't back up into the sink during the drain cycle.

— **Kitchen stove:** Photo #13 shows an **electric range** with 2 ovens. To inspect it, turn the burners on one at a time to see if each one works. Try the bake and broiler elements in each oven to be sure they heat up. Check the lower oven door to see if the spring works and the

---

### SELECTED APPLIANCES

- Dishwasher
- Kitchen stove
- Refrigerator
- Exhaust fan
- Ceiling fan

↖ *Photos #10 and #11*

*Personal Note*
  "One of my inspectors turned on a dishwasher in a brand-new home without checking first. To be honest, there was no one in the home to ask about it. Well, the dishwasher went through its cycle all right, but decided to drain its contents out under the sink and onto the kitchen floor. The discharge hose hadn't been hooked up under the sink.
  "Of course, the inspector cleaned up the mess."
                    *Roy Newcomer*

↖ *Photo #12*

← *Photo #13*

*"I turned on a ceiling fan in a dining room during an inspection. Within seconds, the whole fan, wiring and all, fell from the ceiling. I caught it just before it crashed to the dining table and almost broke my wrists in the process.*

*"Since that experience, I always caution my inspectors to check a ceiling fan for secure attachment __before__ turning it on. Don't assume that everything is all right.*

*"By the way, the customer was so thankful. She could only imagine what would have happened if her family had been sitting at the table when the fan came down."*

Roy Newcomer

*Photo #14* ➔

oven door doesn't fall open. We suggest that you don't get involved with testing the timer, clock, or any lights. These are minor points. CAUTION: *Remember to turn off the burners and ovens when you're finished.*

Some ranges have a **microwave** as the top oven. You can test a microwave by putting a cup of water in it for 60 to 90 seconds.

— **Refrigerator:** This is a fairly cursory inspection, just enough to make sure the refrigerator is operating. Open the freezer door and feel packages in the freezer to see if they're frozen solid. Open the refrigerator door and leave it open until you hear the motor kick in. Check the gaskets around the door.

— **Exhaust fan:** Test exhaust fans for operation. They may be present on the range hood or on an exterior wall. *Be sure that ductwork for any fans exhausts to the exterior.* (You would have noticed its outlet during your exterior inspection.) Examine the condition of the filter in the range hood and see if the grease trap and charcoal pack are missing.

— **Ceiling fan:** To inspect a ceiling fan in a kitchen or any other room, first touch the fan to see if it's securely attached to the ceiling. *Don't turn it on if it feels loose.* Run the fan on high and low speeds to check if it wobbles or vibrates too much in its mounting.

### Inspecting the Bathroom

Again, the home inspector should develop a routine for inspecting the bathroom and then not vary from the routine. It's important to not miss any of the items to be inspected. We've suggested one option here:

• **Installed bathroom lighting:** Try the light switch by the door when you enter the room.

• **Walls, windows, door, and ceiling:** Photo #14 shows a **typical small bathroom**. Before focusing on the toughest bathroom items, inspect the walls and ceiling. If a first-floor bathroom sits directly under a second-floor bathroom, pay careful attention to the ceiling for any signs of leakage from above. As you inspect walls, take the time to test towel racks, toilet paper holders, and tub grips for secure attachment to the wall. Notice if there is a **heat source** in the bathroom. Be sure to report the absence of heat in the bathroom.

Bathroom windows can have water damage because of their proximity to the tub or shower. Sometimes, homeowners have plastic curtains on bathroom windows that, if not opened often enough, can cause sills and sashes to rot and mildew. Be sure to inspect the back of the door as well as the front. Check to see if the door stop is in place to protect the wall behind the door.

- **The floor:** Always inspect the bathroom floor carefully. Look for deterioration of the floor and subfloor around the edges of the room, under the sink, and around the toilet, tub, and shower. Push against flooring with the toe of your shoe or get down on your hands and knees to push against and probe flooring, especially around the toilet and tub. You don't want to miss wood rot and water damage.

With plastic resilient floor coverings, check for deterioration of the sheet or tiles and loose adhesive. With ceramic tile or other types of tile, examine the condition of the grout. Broken or missing grout lets water into the subflooring. Water can also get through the floor if it rises over the sanitary base around the edges. Grout at the edges of the floor isn't enough to stop water, and the sanitary base should be present. If a floor is carpeted, be suspicious that it's covering something wrong underneath. Lift the carpet at a corner, if possible, to check the floor underneath.

- **Outlets:** Test bathroom outlets for polarity and grounding and proper distances from the tub. Check for the presence of a GFCI on outlets within 6' of water and test those that are present. If outlets don't have GFCI's, recommend that they be installed. (See pages 51 to 53 for tests.)

- **The sink:** Check the sink for general condition, including cracks, nicks, and other damage. Photo #15 a **marble bathroom sink** and cabinet. Run the water at the sink and check for the presence of hot water. Check water pressure by turning on the sink and tub faucets and then flushing the

---

**IN THE BATHROOM**

Leaks and water damage, wood rot, and metal shower pans. Don't miss them. Ever. Examine bathrooms from below whenever possible.

← *Photo #15*

---

toilet. Be sure to open the cabinet below the sink and inspect the piping, trap, and floor inside the cabinet.

*Photo #16* ➜

Photo #16 shows a **pedestal sink**. Be sure to examine the piping behind a pedestal-style sink too. (By the way, the date of manufacture is usually stamped on the bottom of a pedestal sink. This can help you determine the age of the home, assuming it's the original sink.)

- **The toilet:** Flush the toilet during your inspection to test its operation. Examine the tank and bowl for any signs of cracks, damage, and leaking. Check to see if the toilet is tightly secured to the floor. Straddle the bowl, hold it between your knees, and rock it back and forth gently to do this. Or you can bend down and gently lift the bowl to test for looseness. A loose bowl should be reported.

- **Tub and shower:** Examine tubs, shower stalls, and tub and shower combinations for the condition of the fixtures and faucets. Look for leaking. Some older showers had a ceramic tile floor laid over a **metal shower pan**, which was made or either lead or tin. These metal shower pans are notorious for leaking, and if it isn't leaking now, it surely will soon. *Never miss reporting the presence of a metal shower pan.*

You should suspect that a metal shower pan is present if the floor of the shower is laid with the very small ceramic tiles, as that was the style with these pans.

If you can't tell if the shower pan is leaking, you can block the shower drain and let water fill the bottom of the shower. Then uncover the drain and try to observe leaking from under the shower, if possible. In any case, always be sure to observe the shower from underneath and look for evidence of leaking. You can often see the corrosion and leaking around the shower drain and water will have done a great deal of damage to the floor and underlying structures.

*Photo #17* ➜

The wall around the tub, or surround as it's called, must be inspected too. Photo #17 shows a **ceramic tile surround** for a bathtub. Ceramic tile used to be applied with concrete mortar pressed against wire mesh, which was called setting it in **mud**. That makes for a heavy wall requiring strong framing. The wall was water resistant as long as the grout lasted and the reinforced wall didn't crack. Tiles can be applied to plywood or gypsum board with water resistant adhesives for a lighter wall.

Check the tile for condition and for broken and missing grout. Ceramic tile surrounds are vulnerable to leaking at the junction of the tub and the tiles where caulking may be inadequate. Water seepage is common around the faucet (and shower tap, if a shower is installed). Use your screwdriver to tap against tiles to test for looseness — loose tile makes a crackling noise, while tight tile rings. Pushing against the tiles that yield can indicate loose tiles or deterioration of the plaster or plywood backing. Inform customers that tile work should be repaired and open joints regrouted to prevent water damage to the wood framing and ceiling below.

Photo #18 shows a **fiberglass surround**. This type of molded surround, applied over old tiles or plywood, is designed to be waterproof. Some units, however, have a horizontal gap between the surround and the tub that needs to be kept well caulked with a flexible compound. Be sure to check the very top edge of the surround too, where water can get in and seep into the wall. These edges are often overlooked because of their height.

NOTE: Doors on showers and tubs should be safety glass or plastic rather than regular glass so the bather can't be injured by falling against them.

- **Exhaust fan:** First, check that the fan has its own switch. You may find some fans wired to the light switch so they go on every time the light does. This is wasteful, and you might want to suggest rewiring it. Turn on the bathroom exhaust fan to see if it works. Noisy fans are caused by dirty or distorted blades. Be sure that fan ductwork exhausts to the exterior.

## Inspecting Other Rooms

Routine is important too in examining the other rooms in the house, and you may approach bedrooms differently from family rooms. Give the walls, ceiling, and floor in each room a good inspection. Remember to look for a heat source in each room

*For Beginning Inspectors*

*Get out your flashlight, GFCI tester, and neon bulb tester again and head for the bathroom for a complete inspection. After doing your own, bother your friends again and practice inspecting as many bathrooms as you can.*

← *Photo #18*

---

### HOMEOWNER TRICKS

- Covering bathroom floor problems with carpeting.

- Not using the shower for a while before the home inspection to hide leaking.

- Misrepresenting the age of the home. (Check for date under toilet tank lid or under pedestal sink.)

- Arranging items on countertops to cover up damage.

- Boarding over closet ceilings to hide water damage.

- The list goes on . . .

and to perform the necessary electrical tests on selected outlets. Operate a representative number of windows, and don't forget to look behind doors in every case. As you conclude your interior inspection, remember to inspect foyers and hallways and any closets in them, including closets under stairs.

During your inspection, check for the presence of **smoke detectors** in the home. Test each one for operation. Advise customers that there should be a smoke detector on each level of the home and each one should be tested once a month.

*Photo #19* ➔ A WORD ABOUT CEILING STAINS: Always report any sign of water damage or leaking in a ceiling. And when you find staining, try to figure out what's causing it. Photo #19 shows some **water stains** on a bedroom ceiling. Just because the stain is faint and not too large, it doesn't mean that you should not investigate further. Is there a bathroom up above? Is a waterbed leaking in an upstairs bedroom. Is the roof leaking? Try to solve the problem.

*Photo #20* ➔ Photo #20 shows **extensive and ongoing staining** along one side of a ceiling. This was a first-floor room with a bedroom above it, so the water damage wasn't from a bathroom. We went back outside to see that side of the house and saw a balcony off the upstairs bedroom. A trip to the bedroom and out onto the balcony revealed wood rot on the balcony at the junction with the house. Water had been seeping into the wall at the junction and leaking into the ceiling below. It took some time to locate the source of the problem, but it was important to do so.

## Reporting Your Findings

Your inspection report should have a separate page to record your findings on the kitchen. Separate pages or sections should be devoted to the bathroom, preferably a section for each bathroom in the house. Smaller sections should be available for each of the other rooms such as the living room and bedrooms.

Here is an overview of how to report your findings on the room-by-room interior inspection:

- **Kitchen:** Report your findings on the condition of the ceiling, walls, and floors in the kitchen. Note defects such as ceiling and wall cracks, sloping and uneven floors, wood rot, and signs of leaking. Note the presence or absence of a heat source. Write your findings on the condition of the countertops and cabinets, noting if any hardware is missing or broken or any structures are loose or detached. Write your remarks on the doors and windows in the kitchen. It's a good idea to report on windows in general as we mentioned on page 48 and also for each room. Record your findings on the plumbing situation in the kitchen. Rate the functional flow and drainage at the sink as adequate or poor. Then, record the results of testing outlets in the kitchen. Remember to classify reverse polarity and an open ground near water as a **safety hazard**. We suggest that you recommend GFCI's near the sink if they're not already in place.

It helps to have a list of appliances in your report with a checkoff list so you can indicate if the appliance was present, was tested, and if it's operating.

- **Bathroom:**  As with the kitchen, record your findings on walls, ceilings, floors, doors, and windows in the bathroom. Note the presence or absence of a heat source. Rate the functional flow and drainage at the bathroom fixtures as adequate or poor. Then, record the results of testing outlets in the bathroom. If a ventilation fan is present, indicate if you've operated it and whether or not it's operating.

It helps in filling in the bathroom report if each fixture is listed on the page with a checkoff list for rating faucets, pipes, fixture condition, and so on. Note defects such as leaking faucets, tub needs caulking, tile needing grout, loose toilet bowl, and so on.

- **Other rooms:**  Begin by identifying which room you're reporting your findings on. Record your findings on the walls, ceiling, floor, windows, doors, heat source, and electrical situation for each room. Don't miss mentioning any water stains from leaks, rotted flooring, and other conditions stated in the Don't Ever Miss list.

---

### DON'T EVER MISS

- Missing power, lighting, or heat source in any room
- Water damage, wood rot, and water stains, old or new
- Ceiling, wall, and floor conditions indicating structural problems
- Loose, cracked, or damaged ceiling and wall facings
- Electrical safety hazards
- Metal shower pans
- Plumbing leaks
- Fans not exhausted outside

# WORKSHEET

*Test yourself on the following questions*
*Answers appear on page 66.*

1. According to most guidelines, the home inspector is required to:

    A. Test all electrical outlets for current.
    B. Test all electrical outlets for reverse polarity.
    C. Test one outlet per room for current.
    D. Test only bedroom outlets for open grounds.

2. According to most guidelines, the home inspector is required to:

    A. Operate a representative number of plumbing fixtures.
    B. Operate all plumbing fixtures.
    C. Test only kitchen faucets for water pressure.
    D. Test only bathroom faucets for water pressure.

3. According to most guidelines, the home inspector is not required to:

    A. Inspect walls, ceilings, and floors for water stains.
    B. Operate all windows.
    C. Inspect both sides of interior doors.
    D. Inspect the inside of kitchen cabinets.

4. Which GFCI's in the home should be tested?

    A. All of them in the home
    B. One in each room that has them
    C. A representative number of them
    D. Only those in the bathroom

5. What should the home inspector do upon finding a metal P-trap under the sink?

    A. Report it as being illegal and suggest it be replaced with an S-trap.
    B. Inspect the trap for corrosion and rust.
    C. Inform the customer there's no venting.

6. Inspect the trap for corrosion and rust. What type of shower construction will almost certainly leak eventually?

    A. One with a fiberglass surround
    B. One with a metal shower pan
    C. One with a shower and tub combination
    D. One with a ceramic tile surround

7. Which area in the home can the home inspector bypass during the interior inspection?

    A. Small rooms
    B. Closets under stairways
    C. Foyers
    D. None of the above

8. What condition should alert the home inspector to a leaking seal in a thermal-pane window?

    A. Condensation on the inside surface
    B. Framing out of square
    C. A rotted sill
    D. Discoloration or cloudiness between the lights

9. What condition could be reported as a major repair?

    A. A loose ceiling fan
    B. Nails pops in the drywall
    C. Plaster ceiling close to collapse
    D. A missing heat source in a bedroom

10. Upon finding a carpeted bathroom floor, the home inspector should:

    A. Report the floor as water damaged.
    B. Report the condition as a homeowner trick.
    C. Try to lift the carpet and examine the floor. Not perform a floor inspection

# Chapter Eight
## THE INSULATION AND VENTILATION INSPECTION

In this chapter, you'll learn about the guidelines for the insulation and ventilation inspection.

### Inspection Guidelines and Overview

These are the guidelines that govern the inspection of the insulation and ventilation systems. Please review them carefully.

**Guide Note**

*Pages 65 to 67 lay out the content and scope of the insulation and ventilation inspection. It's an overview of the inspection, including what to observe, what to describe, and what specific actions not to take. Study them carefully.*

| Insulation and Ventilation Systems | |
| --- | --- |
| OBJECTIVE | To identify major deficiencies in the condition of the insulation and ventilation systems. |
| OBSERVATION | Required to identify and report: <br> • Insulation and vapor retarders in unfinished spaces <br> • Absence of same in unfinished space at conditioned surfaces <br> • Ventilation of attics and foundation areas <br> • Kitchen, bathroom, and laundry venting systems |
| ACTION | Not required to: <br> • Report on concealed insulation and vapor retarders. <br> • Report on venting equipment which is integral with household appliances. <br> • Report R-value of insulation material <br> • Activate thermostatically operated fans |

Most standards of practice provide a good outline of what is to be inspected and what is to be reported during the inspection of the insulation and ventilation in the home. Not all details are included in this chart. These details will be presented in the following chapters.

For the inspection of **unfinished spaces** (attics and crawl spaces, for example), most requirements in the standards of practice for the structural inspection state that the home inspector is required to make every effort to get into and inspect these unfinished spaces. And, for his or her own protection, the home inspector should report on how access was gained to these spaces. Reporting in this way lets customers know that some defects may not be found if access is limited. This is important for the inspection of insulation and ventilation as well as the other aspects of crawl space and attic inspection.

- *The home inspector is required to enter underfloor crawl spaces and attic spaces except when access is obstructed, when entry could damage property, or when dangerous or adverse situations are suspected.*

- *The home inspector is required to report the method used to observe underfloor crawl spaces and attics.*

The insulation and ventilation inspection consists of inspecting the following:

- **Insulation and vapor retarders:** For the most part, insulation and vapor retarders are only visible in unfinished areas such as attics and crawl spaces. The home inspector is expected to examine these areas for insulation and to report its **presence and location**, identify the **type** of insulation, inspect its condition and installation, and determine if the insulation is adequate. The inspector also attempts to determine the presence or absence of a vapor barrier on the warm side of the insulation. Note that the home inspector is **not required to report on concealed insulation and vapor retarders**. This instruction refers to insulation in exterior walls that is not visible (although we'll suggest a method to help you get an idea of exterior wall insulation) and vapor retarders that may be hidden above or below insulation.

Where plastic board insulation is used, it should be covered with a non-flammable material such as 1/2" drywall. The home inspector reports any exposed plastic board insulation as a **fire hazard**.

In unheated areas, the home inspector also checks plumbing pipes, heating ductwork, and venting system piping for the presence of insulation.

- **Attic and foundation ventilation:** The home inspector examines the attic and foundation areas for proper ventilation in the form of roof vents and crawl space perimeter vents. The types of roof vents are identified. The home inspector determines if vents are working, are

*Worksheet Answers (page 64)*

1. C
2. B
3. B
4. A
5. B
6. B
7. D
8. D
9. C
10. C

not blocked by insulation, and are doing their job. Much attention is paid to **condensation** in these areas and **deterioration of structural members**, both of which are the results of poor ventilation. The adequacy of ventilation is based on such findings. Ventilating fans are also examined.

## THE INSPECTION

- Visible insulation and vapor retarders
- Attic and foundation ventilation
- Exhaust venting
- The attic

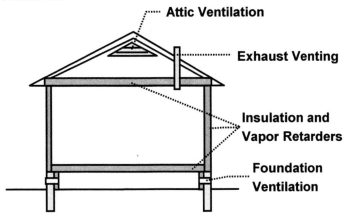

**Attic Ventilation**

**Exhaust Venting**

**Insulation and Vapor Retarders**

**Foundation Ventilation**

- **Venting systems:** Kitchen, bathroom, and laundry venting systems must be vented to the outside. During the insulation and ventilation inspection, the home inspector makes sure that house vents do not exhaust into the attic and contribute to a moisture problem. The home inspector is **not required to report on venting equipment that is integral with household appliances**.

- **The attic:** For convenience, we've included the inspection of the attic along with the insulation and ventilation inspection. This guide presents the attic inspection, which is covered in pieces in our other guides, all together for you. The full attic inspection includes examining the roof and floor structure, the chimney chase, insulation, roof vents, plumbing pipes, exhaust venting, and electrical features. The home inspector examines the attic for any evidence of leaking and/or condensation and determines the adequacy of insulation and ventilation.

# Chapter Nine
# INSULATION

**Guide Note**
Pages 68 to 84 present the study and inspection of insulation in the home.

At one time in our history, little attention was paid to insulation because fuel was inexpensive. The use of insulation in homes started when fuel costs increased. Insulation slows the rate of heat loss from a house. Whenever there is a difference in temperature between the interior and exterior of structures such as walls, air currents form to move heat from the warm side to the cold side of these structures. The tiny fibers or foam bubbles in insulation slow down the movement of these air currents and consequently reduce heat loss.

## Types of Insulation

Different types of insulation have different capacities to resist heat transfer. Insulation is given an **R-value**, which is a number used by the industry to represent the **amount of heat resistance per inch of thickness**. The higher the R-value, the greater the resistance. The U.S. Department of Energy recommends an **R-rating**, or total heat resistance, for insulation in various locations in the home. For example, an R-rating of 38 is recommended in attics in northern climates. A type of insulation with an R-value of 2.9/inch would require a thickness of about 13" to meet an R-rating of 38.

**R-value × number of inches = R-rating**

Some common insulation products are produced in **batts and blankets**. Insulating batts are manufactured in lengths of 4' and 8' and in thicknesses of 2" to 6". Blankets are available in continuous rolls in thickness from 1 1/2" to 3". Widths of both batts and blankets are 15" or 23" to fit between studs and joists at spans of 16" and 24". The insulating wool-like material in batts and blankets is made up of thin manufactured fibers created from glass or mineral waste such as rock and slag:

- **Fiberglass batt and blanket insulation** is made of threads of glass covered with a coating that binds the fibers in place. The insulation may be white or dyed a distinctive color (such as pink) by the manufacturer. Fiberglass insulation in the form of batts and blankets has an R-value of 3.1/inch (requiring 12 1/2" of attic insulation). It is

### TYPES OF INSULATION

- Batts and blankets
- Loose fill
- Rigid board
- Site-foamed

**Definitions**

*The R- value is a number that represents an insulation material's resistance to heat flow per inch of thickness. An R-rating is the total heat resistance for a given thickness of insulation.*

*Batts and blankets are lengths of fibrous insulation manufactured to fit between studs and joists. Batts are pre-cut to length; blankets come in continuous rolls.*

resistant to moisture, mildew, fungus, and vermin. Some types of fiberglass insulation are non-combustible. You'll often see 3 1/2" fiberglass batts in the attic of older homes, installed when the home was built. The paper facing on the insulation stated the R-rating of the insulation as R-11 (3.1 R-value × 3 1/2" thickness). Insulation requirements have changed in the last 20 years. This amount of fiberglass batt insulation wouldn't meet the recommendations for attic insulation today.

- **Rock wool batt and blanket insulation** is made from fibers created by blowing steam through molten rock or mineral waste. Rock wool is typically dark gray in color. In batt and blanket form its R-value is 3.7/inch (requiring only 10 1/2" of attic insulation), slightly better than fiberglass. This type of insulation is resistant to rust and rot and has good resistance to fire.

  Photo #21 at the back of this guide shows grayish **rock wool blanket insulation** laid in an attic. The photo also shows some delamination problems in the sheathing, but we'll talk about that later.

Batt and blanket insulation may be faced on neither, one, or both sides with plain paper, treated kraft paper, or metal foil. **Kraft paper** is an asphalt impregnated paper that is water resistant enough to serve as a vapor barrier. Metal foil on batts and blankets will reflect heat if the foil side faces a free air space at least 3/4" thick. You may be familiar with foil-faced insulation on ductwork and piping, where the insulation faces open space. Of course, if foil facing is covered with drywall, its heat reflective properties are lost.

CAUTION: During installation, when the fibers in batt and blanket insulation can become airborne, the tiny fibers can be a skin and eye irritant or can be breathed in. Glass fibers may be more irritating than mineral fibers, but once in place, there is no hazard. However, the home inspector may loosen glass or mineral fibers if the insulation is handled during an inspection. Some inspectors wear a dust mask and a jump suit for protection.

Another type of insulation is called **loose fill**. Loose fill insulation is normally supplied in bags or bales and can be poured or blown into wall cavities or in horizontal spaces such as between ceiling joists in the attic. Loose fill insulation is

---

| BATTS AND BLANKETS |
| --- |
| • Fiberglass |
| • Rock wool |

← *Photo #21*

*Definitions*

*Rock wool* is an insulating fibrous material made by blowing steam through molten rock or slag.

*Kraft paper* is a water-resistant asphalt impregnated paper that can be used as a vapor barrier.

*Loose fill* insulation is a loose material such as fiberglass, rock wool, or cellulose poured or blown in place between wall studs and attic joists.

*For Beginning Inspectors*

*The best way to learn how to identify different types of insulation products is to see them. Identification of the type of insulation is required during the home inspection. Visit retail outlets to see what's available. Talk to builders to find out what is being used in construction in your area. And take a look at as many attics as you can.*


*For Your Information*

*Find out the recommended R-ratings for your area of the country.*

generally less efficient inch for inch than batt or blanket insulation. Moreover, the insulation in walls can pack down over time due to vibrations of road traffic, wind, and activity in the home, leaving uninsulated space at the top of the wall cavity. Loose fill insulation may be made of the following materials:

- **Fiberglass loose fill insulation** is made of the same glass threads as batt and blanket insulation, but the threads are loose so they can be blown in. The R-value of fiberglass loose fill is 2.2/inch (requiring about 17 1/2" to meet recommendations for attic installations). Fiberglass as loose fill is less efficient than in batts and blankets.

- **Rock wool loose fill** insulation is essentially the same material as used in batts and blankets. However, its R-value is less as a loose fill — only 2.9/inch (requiring 13" in the attic).

- **Cellulose loose fill insulation** can be blown or poured into wall cavities or horizontal attic spaces. It's made of shredded recycled newspaper or wood fibers treated with a fire retardent. Cellulose loose fill is usually gray in color and feels like lint. Its R-value is 3.6/inch (requiring only 10 1/2" in the attic), a higher value than either fiberglass or rock wool. But cellulose has a greater tendency to settle in wall cavities.

The fire retardent chemicals used in treating cellulose are in either granular or liquid form. The fire retardent may not perform over time. Granules can eventually sink to the bottom of the loose fill. Liquid applications remain water soluble and can be leached out if the insulation becomes wet. Both conditions render the insulation vulnerable to fire. That's why the home inspector should carefully inspect electrical conditions in an attic filled with cellulose insulation. Junction boxes must be covered, and recessed lights should not be covered with the insulation.

Cellulose easily absorbs water. If cellulose loose fill in the attic becomes water soaked from condensation or a roof leak, the fire retardent chemicals can have a corrosive effect on the metal gussets of attic trusses and on electrical armored cable and pipes in contact with it.

- **Vermiculite loose fill insulation** is made from mica, which is heated and expanded. As loose fill, its small particles are rectangular in shape. Vermiculite is fireproof and resistant to rot and mildew, although it does absorb moisture. Vermiculite's R-value is 2.1/inch (requiring about 18" in the attic). Another natural loose fill insulation is **perlite**, which is made from volcanic rock. Both of these insulations tend to be more expensive than others.

**Rigid board insulation** can be used on foundation and masonry walls, under vinyl and aluminum siding, and under roofs. It comes in widths of 24" and 48" and can be made from fiberglass, wood fiberboard, or foamed plastics such as polystyrene or urethane. Rigid foamed insulation is often installed on the exterior of a foundation during construction.

- **Fiberglass rigid board**, with an R-value of 4/inch, is commonly used for insulating the exterior of foundation walls. On the exterior it also serves as a drainage layer around a basement.

- **Polystyrene rigid board insulation** is a plastic foam formed into boards. There are two kinds of polystyrene used — **extruded polystyrene**, where the foam is produced by extrusion, and **bead board**, where the foam is produced by expanding and then fusing granular pellets. They have R-values of 3.9/inch and 3.6/inch respectively. Both are used as exterior or interior foundation insulation and under slabs as well as under sidings and on roofs.

Polystyrene can burn easily and rapidly once ignited and melts as it does, spreading fire with the molten flow. Therefore, rigid polystyrene boards used in interior spaces must not be left exposed. They're required to be covered with 1/2" of plaster or drywall for fire safety. Any exposed polystyrene boards found during a home inspection should be reported as a **safety hazard**.

- **Urethane rigid board insulation** is made with a foam that has bubbles filled with chlorofluorocarbon gas and is yellowish orange in color. Its R-value is high at 6/inch and provides good insulation. However, urethane boards also pose a fire hazard if they're left exposed in a home, and they give off toxic fumes in a fire.

---

### RIGID INSULATION
- Fiberglass
- Polystyrene, extruded or bead board
- Urethane

---

***Definitions***

*Cellulose is a loose fill insulation made of recycled newspaper or wood fibers treated with a fire retardent.*

*Vermiculite is a loose fill insulation made from mica.*

*Polystyrene and urethane are plastic foams that can be formed into rigid boards used in insulating foundations, exterior walls, and roofs. Urethane is also available in less rigid forms and as a foamed-in-place insulation.*

```
┌─────────────────────────────┐
│      SITE-FOAMED            │
│                             │
│  • UFFI                     │
│                             │
│  • Urethane                 │
│                             │
│  • Airkrete                 │
│                             │
└─────────────────────────────┘
```

Urethane foams can be produced in less rigid forms from limp to spongy. This softer urethane insulation may be seen in a home to seal openings between the foundation and sill or at wall penetrations. It is recognizable by its yellow-orange color and shiny, wet appearance.

Another type of insulation used is called **site-foamed insulation** or foamed-in-place insulation. As the name implies, this type of insulation is produced at the site. The insulating material in a syrup form is mixed with a reactant creating a foam that can be poured or sprayed into structural cavities.

- **Urea formaldehyde foam insulation** (abbreviated UFFI), used in the 1970's and early 1980's, was produced onsite from a syrup mixed with formaldehyde as the reactant. It was discovered that UFFI released formaldehyde gas into the home during the first year after installation and under high temperatures and moist conditions. The foam itself is a good insulator, identifiable by its light color, soft foamy appearance, frail crumbly structure, and habit of squeezing out of openings like soap lather. UFFI is also resistant to fire, rot, and fungus.

  UFFI is suspected of being a **health hazard**. The release of formaldehyde gas in concentrations of more than 5 parts per million (ppm) can cause coughing, sneezing, and chest constriction in sensitive people. From 50 to 100 ppm can cause pulmonary edema, pneumonitis, and death. The American Cancer Society considers formaldehyde a potential carcinogen. The use of UFFI was banned in 1983, although the government reversed itself a few years later. But UFFI is still not installed today.

  If the home inspector finds or suspects old UFFI insulation in a home built in the 1970's or early 1980's, it is likely that it has already released its formaldehyde. The customer should be informed of the finding and told that its presence is most likely no longer a health concern.

- **Urethane foam**, as described on page 71, can also be mixed on site, then poured or sprayed to foam in place. Again, as a plastic foam, urethane site-foamed insulation must be covered with plaster or drywall.

*Definitions*

*Site foamed insulation, made of various syrups and reactants, is mixed on the site and foamed in place.*

*UFFI (urea formaldehyde foam insulation) is a site-foamed insulation used in the 1970's and early 1980's, but not used after 1983. UFFI releases formaldehyde gas into the home as part of the evaporation process and under conditions of high temperatures and humidity.*

*Airkrete is a site-foamed insulation made of a mixture of a cement containing syrup and air.*

- **Airkrete** is a form of site-foamed insulation made of a cement containing syrup that is foamed with air and poured into cavities as a non-flexible insulator. It provides a good alternative to the plastic foams as it doesn't burn or produce toxic fumes. However, it is not water resistant and may cause structural rot if poured into wall cavities.

## Asbestos in Insulation

Asbestos is a mineral fiber found in rocks. There are different kinds of asbestos fibers, all of which are fire resistant and not easily destroyed or degraded by natural processes. The mere presence of asbestos in building materials and home products is not necessarily a **health risk**. The danger occurs when asbestos fibers are released from the material to the air and breathed in. Once inhaled, asbestos fibers can become lodged in tissue for a long time, and cancer of the lung or stomach can develop after many years of exposure. Experts say that no level of exposure to asbestos fibers is totally safe.

The **Environmental Protection Agency** (EPA) has placed restrictions on the use of asbestos in home products since the 1970's and ordered a total phase-out of its use in 1989. For asbestos already in the home, the EPA recommends either removal of the asbestos product by a qualified asbestos-removal contractor to an approved disposal site or encapsulation of any asbestos left in place in the home. The process of removal is dangerous since that's when the fibers are disturbed and can be released to the air.

Asbestos is used in products because of its strength, its good thermal and acoustic insulating properties, its use as a binder, and its fire resistance. It was used in the many materials including asbestos cement siding and roofing, vinyl floor coverings, textured paints, and patching compounds. It can still be found in homes built before the EPA began ruling on the issue in the 1970's. Asbestos in insulation includes the following:

- **Wall and attic insulation:** Generally, asbestos fibers were used as loose fill insulation in homes built from the 1930's through the 1950's. The fill is likely to look fibrous or powdery. Some rock wool insulation contained asbestos fibers. It's difficult to know for certain whether insulation

---

**ASBESTOS INSULATION**

Be careful when you are inspecting any insulation you suspect contains asbestos. Don't disturb it. It's dangerous when asbestos fibers are released into the air and breathed in.

contains asbestos or not, and the home inspector should not engage in guessing. If asbestos is suspected, it should be reported.

- **Water pipe and heating duct insulation:** A common pipe and duct insulation that contains asbestos looks like corrugated cardboard when viewed from the end. This insulation, if deteriorating or damaged, is one of the most common causes of releasing asbestos fibers into the home. Pipe and duct insulation can be left in place if encapsulated first with plastic wrap and then duct tape. Removal by qualified disposal technicians is expensive.

*Photo #22* →

Photo #22 shows **water pipes wrapped in asbestos insulation**. Look at the corrugations in the insulation. You can be almost certain that the insulation contains asbestos because of them. Note that this insulation is in poor condition and is coming apart at the seams. For health reasons, the insulation must be taken care of either by removal or encapsulation.

- **Plaster reinforced with asbestos:** Plaster mixed with asbestos was used to insulate boiler and furnace parts and supply ducts, especially at elbows. A plaster covering was used on the bonnets of old gravity warm air furnaces. When you see it, you can be almost certain that asbestos is present, based on our knowledge of the practices used years ago.

*Photos #23 and #24* →

Photo #23 shows **a boiler** with all the pipes wrapped in the asbestos-plaster covering. Photo #24 shows **a heat duct elbow** going upstairs which is covered with an asbestos-like material. Notice the material's condition where it's coming off the sheet metal. This situation needs to be remedied. The home inspector can recommend that the asbestos-like tape be painted over with latex paint or removed.

- **Millboard:** Made of asbestos and gypsum, millboard can be found as a protective insulator on the walls and floors near a wood stove. Millboard was often mounted inside boiler and furnace cabinets as well. Exposed millboard should be covered with sheet metal to prevent fibers from being released.

Tell customers that the identification of asbestos in the home is beyond the scope of the home inspection and that home inspectors are not trained to identify asbestos. *You're really not trained!* You can make a pretty good guess with pipe and duct wrapping and plaster with asbestos in it, but you can never be certain. Positive identification can only be made by sending samples to a qualified laboratory for testing. Inform the customer whenever asbestos insulation is *suspected* in the home. Even when you're sure you've found asbestos insulation, it's best to report that although you're *fairly certain* it contains asbestos, a sample be sent to a lab for confirmation.

## Insulating the Home

Insulation should be installed around the heated living space in the home. Insulation should be installed with an **air/vapor barrier** against the warm side of the living space. Spaces on the cold side of insulation should be well ventilated. The vapor barrier is a thin sheet of material such as polyethylene film, aluminum, or kraft paper which is moisture resistant. This barrier may be a separate sheet, but some insulations have a vapor barrier facing. The purpose of the barrier is to prevent structural damage due to the condensation of moisture from the home that would normally travel from the warm to cold sides of the ceiling below the attic, floors above unheated spaces, and exterior walls. The improper installation of insulation and the vapor barrier can actually promote condensation and result in structural damage.

The areas shown below should be insulated.

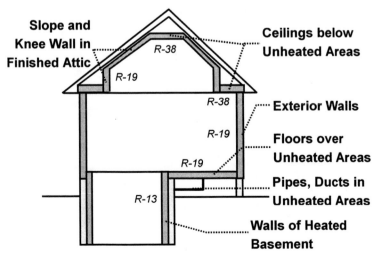

- **Floors over unheated areas:** If the basement or crawl space is not heated, then the floor above that area should be insulated with the vapor barrier against the floor as shown below. An R-rating of R-19 is recommended. Floors above a porch or over an unheated garage should also be insulated. It's not unusual to find crawl space insulation installed with the vapor barrier down, but that's not correct.

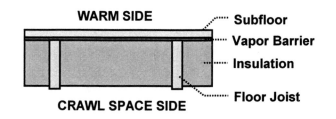

- **Ceilings below unheated areas:** The floor of an unheated attic should be insulated, not the attic walls and rafters. This is not always understood by homeowners who insulate the rafters, thinking an unfinished attic should be warm. The vapor barrier should be installed against the ceiling or warm side of the space.

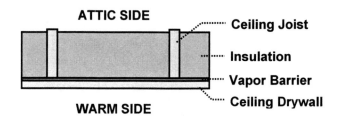

The home inspector may note improper installation in the attic such as faced insulation installed barrier-side-up. And there should be only one vapor barrier. Sometimes, faced insulation is added over existing insulation, adding a second vapor barrier. If 2 layers exist, each with a vapor barrier, the top barrier should be slit open to allow moisture movement. Recommendations are R-38 of insulation in northern and mountainous states and R-30 in the south.

- **The roof slope and knee walls:** If the attic is a heated, finished space, then insulation should be installed on the back of the knee wall and on the roof slope. (When the roof slope is insulated, space must be left behind the insulation for ventilation.) An R-rating of R-19 is recommended.

A **flat roof** above a heated space and the **cathedral roof** should both have the required layer of insulation  The installation of an air/vapor barrier on the warm side of the insulation and the proper ventilation between the insulation and roof surface is essential in both cases.  Both roofs are subject to rot if ventilation is inadequate.

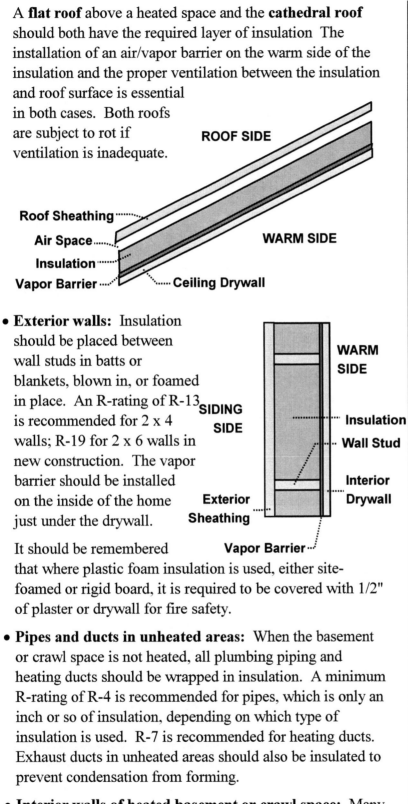

ROOF SIDE

WARM SIDE

Roof Sheathing

Air Space

Insulation

Vapor Barrier

Ceiling Drywall

• **Exterior walls:**  Insulation should be placed between wall studs in batts or blankets, blown in, or foamed in place.  An R-rating of R-13 is recommended for 2 x 4 walls; R-19 for 2 x 6 walls in new construction.  The vapor barrier should be installed on the inside of the home just under the drywall.

SIDING SIDE

WARM SIDE

Insulation

Wall Stud

Interior Drywall

Exterior Sheathing

Vapor Barrier

It should be remembered that where plastic foam insulation is used, either site-foamed or rigid board, it is required to be covered with 1/2" of plaster or drywall for fire safety.

• **Pipes and ducts in unheated areas:**  When the basement or crawl space is not heated, all plumbing piping and heating ducts should be wrapped in insulation.  A minimum R-rating of R-4 is recommended for pipes, which is only an inch or so of insulation, depending on which type of insulation is used.  R-7 is recommended for heating ducts.  Exhaust ducts in unheated areas should also be insulated to prevent condensation from forming.

• **Interior walls of heated basement or crawl space:**  Many areas of the country do not require insulation on the walls of heated basements. If insulation is installed on interior basement walls in northern climates, it need only be applied from the subfloor down to 2' below grade. Rigid board insulation may be applied directly to the walls (but must be covered with drywall) or wall studs can be provided to fill with batt insulation and finished with drywall if desired. If the crawl space is to be heated, then insulation should be installed on the crawl space walls. Insulation in these areas is about R-13.

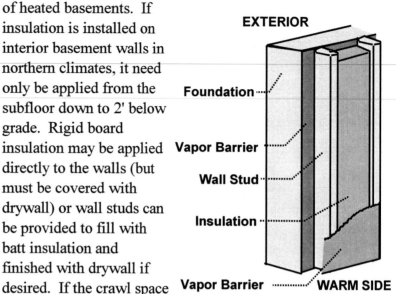

- **Exterior basement walls:** The exterior of the basement may be insulated with rigid board insulation that acts not only as an insulator but to drain water to a perimeter drain tile system. Insulation may extend down the entire wall or to 2' below grade.

In summary, these are the R-ratings currently recommended in new construction by the U.S. Department of Energy for the following areas of the home:

- **R-4:** For plumbing piping in unheated areas
- **R-7:** For heating ducts in unheated areas
- **R-13:** For 2 x 4 exterior walls and crawl space and basement walls, if crawl space or basement are heated
- **R-19:** For 2 x 6 exterior walls, floors over unheated spaces, and knee walls and roof slopes
- **R-30:** For ceilings below attics, generally for most southern states
- **R-38:** For ceilings below attics, generally for most northern and mountainous states

The chart on the next page shows the R-values of various types of insulation covered in this guide, along with the thickness required to meet recommended R-ratings. For example, for fiberglass blankets to meet a recommended R-rating of R-19 for floors over unheated spaces, a 6" thickness is required.

*For Beginning Inspectors*

*It should be obvious that a home inspector needs to measure the thickness of insulation in order to determine its R-rating. Find a ruler or tape measure and inspect whatever insulation is visible in your home. Identify the type of insulation, then multiply its R-value by the number of inches to get its total R-rating. Compare your findings to the latest recommendations as listed here. Perhaps more insulation should be added?*

| R-Values and Amount of Insulation Required | | | | | |
|---|---|---|---|---|---|
| **Insulation Types** | **R-Value** | **R-13** | **R-19** | **R-30** | **R-38** |
| Batts, Blankets | | | | | |
|   Fiberglass | 3.1/inch | 4" | 6" | 9.5" | 12.5" |
|   Rock wool | 3.7/inch | 3.5" | 5" | 8" | 10.5" |
| Loose Fill | | | | | |
|   Fiberglass | 2.2/inch | 6" | 8.5" | 13.5" | 17.5" |
|   Rock wool | 2.9/inch | 4.5" | 6.5" | 10.5" | 13" |
|   Cellulose | 3.6/inch | 3.5" | 5.5" | 8.5" | 10.5" |
|   Vermiculite | 2.1/inch | 6" | 9" | 14.5" | 18" |
| Rigid Board | | | | | |
|   Fiberglass | 4/inch | 3" | 5" | 7.5" | 9.5" |
|   Polystyrene | | | | | |
|     Extruded | 3.9/inch | 3.5" | 5" | 7.5" | 9.5" |
|     Bead board | 3.6/inch | 3.5" | 5.5" | 8.5" | 10.5" |
|   Urethane | 6/inch | 2" | 3" | 5" | 6.5" |
| Site-Foamed | | | | | |
|   UFFI | 4.2/inch | 3" | 4.5" | 7" | 9" |
|   Urethane | 6/inch | 2" | 3" | 5" | 6.5" |
|   Airkrete | 4/inch | 3" | 5" | 7.5" | 9.5" |

*Amounts of insulation in chart are rounded to the nearest half inch.*

## Inspecting the Insulation

The home inspector is required to inspect visible insulation within the home, making every effort to gain access to the unfinished areas (attic and crawl spaces) of the home to do so. During the insulation inspection, the inspector's job is to:

- Identify the type of insulation found.
- Measure the insulation thickness, and give an opinion on its adequacy (optional, but helpful).
- Inspect the condition of the insulation.
- Note any deficiencies in the installation of the insulation.
- Confirm the presence of a vapor barrier, if possible.

During the insulation inspection, the home inspector should watch for the following conditions:

- **Inadequate amount or missing insulation:** Checking the adequacy of the insulation in the attic and other unfinished areas requires measuring it. Inform customers that additional insulation can be installed over the old. The R-

> **INSULATION INSPECTION**
>
> - Identify the type found.
> - Note the average thickness.
> - Note its condition and installation.
> - Locate the vapor barrier.

ratings of existing and new insulation can be added — for example, if 6" of existing fiberglass batt insulation in the attic has an R-rating of R-19 and another 6" is added, then the total R-rating would be R-38 (19 + 19). Remind customers that the new insulation should be unfaced.

The home inspector is not required to report on **concealed insulation**, and your customer should be made to understand this. But try to look at the insulation. If the attic floor is covered, look for any loose boards that can be lifted to see the insulation underneath. You can also try to examine the insulation in an **exterior wall** by removing an electrical outlet or light switch cover plate and looking into the space with a flashlight. Or use a wooden probe like a chop stick or pencil to probe into the area, but be careful when working around electricity. You may not be able to see any insulation in the wall, but don't report it missing if you don't know. Sometimes, the electrician pulled the insulation away from the outlet and you just can't see it. Make sure customers understand.

Examine plumbing pipes and heating ducts in unheated areas such as the attic and crawl space. They should be insulated.

- **Missing vapor barrier:** When you measure the insulation, try to determine if a vapor barrier is present. Identify the type of vapor barrier present, whether kraft paper or a plastic film, and see if it faces the right way (against the warm side). If you can't determine if the vapor barrier is there, report it as **not visible**.

- **Improper installation:** Check insulation for problem installation such as the following examples:

*Photo #25* →

— **Vapor barrier reversed.** Photo #25 in this guide shows **attic insulation installed backwards** with the vapor barrier up instead of down against the warm ceiling. It's also common to find the insulation in the floor above a crawl space with the vapor barrier on the crawl space side rather than against the warm floor.

— **Between attic rafters.** When the attic is intended to be an unheated area, homeowners often mistakenly insulate between the rafters instead of the floor. This lets heats escape from the house into the attic. In a finished attic, look behind the knee wall, if possible, to determine how insulation is installed. It should be located on the attic floor and on the back side of the knee wall, not between the rafters.

— **Gaps.** Watch for gaps in installation. A considerable amount of the insulation's effectiveness is lost when it doesn't completely cover a surface. Gaps may be left around openings for plumbing pipes & exhaust vents in the attic floor.

— **Too close to recessed lights.** When recessed ceiling fixtures are present on the top floor of the house, they should not be covered with insulation in the attic. A clearance of 3" is required between the recessed lights and attic insulation. (Some newer lights may be rated for "insulated ceiling" meaning that they can be covered.) If you come across recessed light buried in the insulation, you don't have to dig through the insulation to determine if the light is rated to be covered. However, the customer should be warned that the owner or customer should do it at a later date as this situation could represent a **safety hazard**.

— **Blocking vents.** Insulation should never block vents. Take another look at **Photo #21** in this guide where some delamination is present on the roof sheathing above the eave. This rock wool insulation has been laid over the eaves where it is blocking the soffit vents, which is interfering with proper ventilation.

Photo #26 shows another example of **blocked soffits**. In this case, we found all the soffit vents blocked so the attic had virtually no ventilation at all. You can see the result of the lack of ventilation in the gross delamination of the roof sheathing.

• **Torn, loose, or damaged insulation:** Watch for any damage to insulation. Insulation in the floor over unheated areas can become loose and begin to fall. Insulation on pipes can be damaged or torn as seen in **Photo #22** in this guide. These cases should be reported. Damaged

---

## INSTALLATION PROBLEMS

- Vapor barrier reversed
- Between rafters
- Gaps
- Too close to recessed lights
- Blocking vents

---

*Personal Note*

   *"It's fairly common to find insulation improperly installed when there's an attic bedroom. Try to get a look behind the knee wall — there's usually a door to a storage area somewhere in the room. It's not uncommon to find the insulation between the rafters behind the wall. I always suggest that the rafter insulation be removed and used on the back of the knee wall and on the floor."*

                    *Roy Newcomer*

← *Photo #26*

insulation no longer does its job of providing heat resistance.

*Photo #27* → Photo #27 shows **falling crawl space insulation**. In this case, the crawl space floor didn't have a vapor barrier, allowing too much moisture in the crawl space. This insulation has been damaged from the moisture.

- **Damp or wet insulation:** Check insulation for dampness and wetness, especially in the attic. A localized wet spot can be an indication of a roof leak. If the insulation is consistently damp, there's most likely a problem with the ventilation in the attic.

*Photo #28* →
- **Presence of asbestos:** As noted on pages 73 to 75, you may find insulation containing asbestos. Review **Photos #22, #23,** and **#24** for examples of asbestos-containing insulation commonly found. Photo #28 shows another example of **asbestos insulation** stuffed around a pipe at the center of the photo. This insulation is wet, falling apart, and should be removed. Remember to report any suspected asbestos in insulation. Remind customers that a sample can be sent to a laboratory for confirmation.

- **Exposed plastic board insulation:** Report any exposed polystyrene or urethane insulation as a **safety hazard**. Be on the lookout for it in garages. Inform customers that such insulation should be covered with 1/2" of protective plaster or drywall.

## Reporting Your Findings

When you're inspecting the insulation, you must make every effort to investigate the attic and other unheated areas where it's possible to gain access. Of course, your customer isn't going to be in these areas with you. When you come out of these spaces, communicate your findings with that customer. Talk over what you were looking for and explain what you found. Since the customer hasn't seen the area firsthand, take your time in describing the following:

- **What you were inspecting** — the insulation on the floor, ceiling, walls, pipes, ducts, or whatever.

***Personal Note***

*"You can give insulation an R-rating based on its R-value and thickness. But you should let customers know that the R-rating is meaningful only if the insulation is intact. When insulation is damaged or falling down, it's not going to meet that R-rating. I always let my customer know that the R-rating works only when the insulation is in good shape."*

*Roy Newcomer*

- **What you were looking for** — improper installation, any damage to the installation, presence of asbestos, and so on.

- **What you were doing** — seeing what type of insulation was present, measuring the insulation, looking for the vapor barrier, and so on.

- **What you found** — inadequate amount of insulation, blocked soffit vents, double vapor barrier, and so on.

- **Suggestions about dealing with the findings** — taking insulation from between the rafters and putting on the attic floor, slicing open vapor barrier of added insulation, wrapping asbestos insulation on pipes, and so on. But with this caution — don't make uneducated guesses about how to remedy any situation.

### Filling in Your Report

Record the findings of your insulation inspection in your inspection report according to the layout of the report. It would be helpful to record these findings on separate pages for each area you've inspected as suggested here:

- **Attic:** First, identify the type of insulation you've found such as fiberglass, cellulose, rock wool, and so on. It's also a good idea to record where the insulation was installed in the attic — between the rafters, on the walls, or on the floor. Report on the condition of the insulation, noting any problems with installation, blocked vents, the presence of old UFFI insulation, and so on. Don't miss reporting buried recessed light fixtures as a **safety hazard**

  We suggest that you record the **thickness of the insulation** and rate it according to your own system of rating. We use a rating system where 0" to 3" is graded *poor* and 10" or more is graded *best*, with gradations in between. You might want to use *adequate* or *not adequate*. We also suggest that you write the R-rating of the insulation. When you review this information with your customer, you can explain what the R-rating finding is and what it should be in the attic (R-30 or R-38). You can also say how much insulation, if any, should be added to meet the recommended rating.

**COMMUNICATING**

If the customer hasn't seen the insulation in the attic or crawl space, talk about what you saw in there. <u>You are their eyes</u>, so take the time to describe your findings. And make sure the customer understands.

**DON'T EVER MISS**

- Improperly installed insulation

- Missing vapor barriers

- Exposed plastic insulation

- Suspected asbestos

- Buried recessed light fixtures

Note whether or not a vapor barrier is present and identify the type found. Note whether it's installed properly.

- **Crawl space:** First, note if insulation is present. If it is, identify the type of insulation found and where it's installed (under the floor structure or on the walls). Record its thickness and R-rating. Note any defects such as improper installation. Note the presence or absence of a vapor barrier. Regarding heating ducts and plumbing pipes, note whether they should be insulated and whether insulation is present. Don't miss reporting suspected asbestos insulation on plumbing pipes.

- **Exterior walls:** If you've had the opportunity to inspect wall insulation, record your findings in your inspection report. You might want to write general comments such as "Wall insulation appears adequate" or "Could not confirm wall insulation during inspection."

- **Garage:** Again, identify the type of insulation present. Be sure to note exposed plastic rigid board insulation as a safety hazard in your report.

# WORKSHEET

*Test yourself on the following questions. Answers appear on page 86.*

1. Which of the following actions is the home inspector <u>not</u> required to perform during the insulation and ventilation inspection?

   A. Observe insulation and vapor retarders in unfinished spaces.
   B. Describe insulation and vapor retarders in unfinished spaces.
   C. Observe ventilation of attics and foundation areas.
   D. Report on concealed insulation and vapor retarders.

2. What is an insulation's R-value?

   A. The thickness of the insulation
   B. Its resistance to heat flow per inch of thickness
   C. The amount of insulation recommended by the U.S. Department of Energy
   D. 3.1/inch

3. How do you determine the R-rating of installed insulation?

   A. Add its R-value and its thickness.
   B. Subtract its R-value from its thickness.
   C. Multiply its R-value and its thickness.
   D. Divide its R-value by its thickness.

4. What type of insulation is shown in Photo #21 at the back of this guide?

   A. Fiberglass batts
   B. Rock wool blankets
   C. Vermiculite loose fill
   D. Cellulose blankets

5. How does the insulating ability of fiberglass loose fill insulation compare with that of fiberglass batts or blankets?

   A. Loose fill is less efficient.
   B. Loose fill is more efficient.

6. Which rigid board insulation has the highest R-value?

   A. Fiberglass
   B. Urethane
   C. Extruded polystyrene
   D. Bead board

7. Which of the following statements is <u>false</u>?

   A. Vapor barrier should be installed against the warm side of the insulated area.
   B. The back of knee walls and the floors behind them should be insulated.
   C. Space should be left behind insulation for ventilation in insulated roof slopes.
   D. Attic rafters should always be insulated.

8. Which insulation is <u>most likely</u> to contain asbestos?

   A. Any insulation in homes built in the 1970's
   B. Exterior foundation wall insulation
   C. Plaster covering on an old gravity warm air furnace bonnet
   D. New insulations made available after 1989

9. Which photo in this guide shows a condition that poses a health risk?

   A. Photo #21
   B. Photo #22
   C. Photo #25
   D. Photo #27

10. Which of the following conditions poses a safety hazard?

   A. Exposed plastic board insulation
   B. Presence of UFFI insulation
   C. Moisture-laden fiberglass insulation in a crawl space
   D. Blocked soffit vents

# Chapter Ten
# VENTILATION

The purpose of ventilation is often misunderstood by homeowners who might think it only has to do with cooling the home. The true purpose of the ventilation process is **to remove moist air from the home**. With unlimited available moisture in a closed space, wood becomes saturated, causing it to rot and deteriorate. Ventilation, by allowing air to move in and out of the structure, replaces damp air with dry air to prevent this problem.

## Crawl Space Ventilation

When the soil under a crawl space is not covered, it continually releases moisture. It is recommended that the dirt or gravel crawl space floor be covered with a **vapor barrier** to prevent moisture from the soil being released into the crawl space. Vapor barriers can be polyethylene sheeting, roofing paper, blacktopping, or concrete. Even with a vapor barrier, the crawl space air contains moisture which can be released into the home. Good ventilation can prevent this from happening.

**Perimeter vents** are required in a crawl space to keep the moisture from rising into the home. Ideally, there should be vents in all four walls, or at least in two opposite walls, so there can be cross ventilation. The amount of venting depends on the size of the crawl space and whether it has a vapor barrier on the floor.

- With a vapor barrier, 1 square foot of free vent is needed for every 1500 square feet of floor area.

- Without a vapor barrier, 1 square foot of free vent is needed for every 150 to 500 square feet of floor area.

*Guide Note*
*Pages 86 to 95 present the study and inspection of the home's ventilation.*

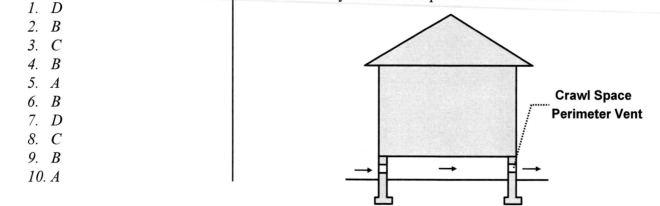

Crawl Space Perimeter Vent

The rules about when to keep crawl space vents open and when to close them are not exact. That's because of differences in climate, location, the type of house construction, whether the crawl space is heated, and whether the crawl space is an extension of the basement. There is some agreement about closing vents in winter, but debate about whether they should be open or closed in summer. The home inspector is responsible only for noting the presence of proper vents in the crawl space, not for their use by homeowners.

Moisture can enter the crawl space in another way. Warm air from the house will naturally migrate to the cooler crawl space. That's why the floor over the crawl space is typically insulated. A vapor barrier should be laid against the warm side of the floor with the insulation on the crawl space side. The vapor barrier can't stop all the moist warm air from entering the crawl space. The cold side of the insulation needs to be ventilated to remove the moist air that does get through. The perimeter vent system provides the ventilation needed to remove moisture collecting in the insulation.

The results of poor crawl space ventilation can be spotted in the rotting of the structural members including the sill, wooden columns and joists, and the subfloor. We've seen homes where the subfloor above a crawl space has rotted away. In one case, the ventilation was so poor that the sill had completely rotted and the house had settled down to the foundation. That house had an uncovered soil floor and no perimeter venting was present.

## A Word about Radon

The home inspector may come across special ventilation techniques in the crawl space or basement or other methods used to remove radon from a home. Radon is a gas that occurs naturally when uranium in soil and rocks breaks down. Because air pressure in the home is usually lower than the pressure in the soil, a house can act like a vacuum and draw radon into the home. Radon can also be found in well water and can be released into the home through running water.

Radon is a radioactive gas that can cause lung cancer. It is estimated that 1 out of 15 homes in the country have elevated radon levels. The EPA and the Surgeon General recommend that

---

**CRAWL SPACE VENTS**

- One on each wall or at least on opposite walls

- With vapor barrier on floor, 1 sq. ft. of free vent for every 1500 sq. ft. of floor area

- Without vapor barrier, 1 sq. ft. of free vent for every 150 to 500 sq. ft. of floor area

## Definitions

*Radon is a gas occurring naturally from the breakdown of uranium in the soil, rocks, and water. The abbreviation pCi/L is a measurement which stands for picocuries of gas per liter of air — the way radon is measured.*

homes be tested for radon levels. It is further recommended that homes with a radon reading of **4 or more pCi/L** (picocuries of gas per liter of air) use a radon reduction technique to reduce that level. Homeowners getting a reading between 2 and 4 pCi/L should consider using a radon reduction technique.

For homes with crawl spaces, the radon reduction technique that may be used is to ventilate the crawl space with or without the use of fans. Another approach is **submembrane depressurization**, which involves covering the soil in the crawl space with a heavy plastic sheet and using a vent pipe and fan to draw the radon from under the sheet. Other techniques are available with basement or slab-on-grade homes, costing from $500 to $2500 to remove radon which enters the home through the soil. Methods used when water is the source of radon in the home are more expensive, generally from $1000 to $4500.

A general home inspection does not include inspecting the home for radon. However, some home inspectors must become qualified (and licensed in some states) to add radon inspection to the list of services they offer. Radon testing is done by placing special canisters in the home for a period of time. These canisters, which absorb radon from the air, must be sent to a qualified laboratory where the radon concentrations can be determined.

## Attic Ventilation

The attic area also requires ventilation. Warm moist air from the house naturally tries to move into the attic. The air/vapor barrier laid along with the insulation on the attic floor plays a large role in preventing warm moist air from entering the attic, but it can't stop it completely. The moist air that does get into the attic must be removed to prevent moisture buildup in the attic. This is done through ventilation.

**Attic vents** are required in the attic to release moisture from the area. Generally, 1 square foot of free vent is required for every 300 square feet of floor area. This requirement is often not met. It should be noted that louvers on vents reduce air flow by 50%, so that when louvers are present, twice the vent area is required. The vent area should be split for air movement in and out. A single vent will do little to move the air.

**Gable vents**, located at the gable ends of the roof, are most effective when the wind is blowing parallel to the roof ridge so that air can move directly through the attic from end to end. They are less effective when the wind is blowing perpendicular to the ridge. In that case, smaller air currents can occur at each end of the attic without a direct pass through the attic.

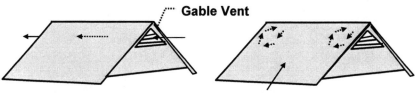

**Wind Parallel**  **Wind Perpendicular**

Gable vents should be unobstructed, and louvers undamaged. Some homeowners cover the gable vents during the winter months in the mistaken belief that this will save heat in the house. Covering the gable vents in the winter is far more likely to allow moisture buildup in the attic and cause damage to the structural members. When inspecting the attic, keep a lookout for gable covers or plastic sheets near the gable vents. Advise customers not to cover the vents in the winter.

The home may have **soffit vents** located in the soffit under the eaves along each side of the roof. Air movement flows into the soffit vents, through the attic, and out. When soffit vents are present, care must be taken to prevent the attic insulation

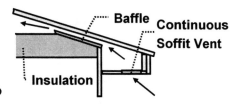

from blocking the air passages at the lower edge of the rafters. A baffle can be put in place to hold the insulation away from the passage area. Some attics have only soffit vents.

The **ridge vent**, which runs the length of the roof ridge, has an open vent area on its underside to move air from the attic.

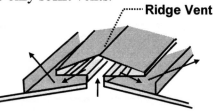

The most effective method of ventilating the attic is the **combination of the ridge vent and soffit vents**. This allows air

*Personal Note*

*"Few homeowners really understand the principles of attic ventilation. I've seen many cases where the owner's actions contribute to moisture buildup in the attic. Blocking vents and covering them in winter to save heat are among the top mistakes. I like to take the time to teach customers to be concerned about good ventilation first and heat loss second."*

*Roy Newcomer*

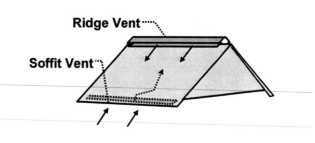

Ridge Vent

Soffit Vent

---

## ATTIC VENTILATION

- Gable vents
- Soffit vents
- Ridge vents
- Roof-top vents
- Turbine vents
- Power ventilator

*Photos #29 and #30* ↗

to enter from the soffit area and move through the attic and up to the roof ridge, providing good ventilation.

A home may have vents on the roof for releasing air from the attic. Photo #29 in this guide shows some **roof-top vents**. These vents are usually round or square metal vents. The ones in this photo, by the way, are improperly flashed — tar should not be used as a flashing method. Photo #30 shows a close up of a **damaged roof-top vent**, which happens to have a proper flashing. These vents can become damaged from storms or trees. The home inspector should be sure to examine them during the roof inspection.

A **turbine vent**, located on the roof face, is an air powered series of vanes on a central rotating spindle. Turbine vents work only when the wind is blowing. On still days, they do no more good than a regular roof vent.

Some attics have **a power ventilator**, which can be located either on the gable end of the attic or on the roof surface between the roof rafters. These electrically powered ventilators may be controlled by a manual switch or by a thermostat. Usually the thermostat is set to activate the ventilator fan when the attic temperature reaches about 100°. In general, the power ventilator is installed for summer use only, when the ventilator's actions remove hot air from the attic and reduce the heat load. It is possible to add a humidistat which controls the fan based on the humidity in the attic. This is very effective in cold climates.

The home inspector should try to operate the power ventilator during the inspection. Check to see that a gable-mounted ventilator doesn't cut down on the free vent space required for the attic. And be sure there is sufficient vent area in the whole attic so the ventilator is not starved for air during its operation.

The home inspector may see a large fan mounted in the upper hall ceiling. This is a **whole house fan** with an intake from the house, designed to change house air every minute or so. A whole house fan may also be mounted in the gable, where house air will be drawn up into the attic through a self-closing louver in the upper hall ceiling. The whole house fan needs to have an appropriate free area for both intake and exhaust. The whole house fan is not a power ventilator.

A homeowner may take precautions with ventilation and do everything possible to eliminate warm moist air from the attic, but then make mistakes about other **venting systems** in the home. Kitchen and bathroom exhaust fans and the plumbing vent stack should vent to the outside. Often, the exhaust venting passes through the attic. There may be deterioration of the piping or open joints that allow warm air to escape into the attic, contributing to moisture buildup.

In worst cases, home venting systems exhaust directly into the attic. This should never be done. The home inspector should report any exhaust vents terminating in the attic and advise customers to remedy the situation. It's not uncommon to see a bathroom vent exhausting into the attic. When that occurs, the extra moisture thrown into the attic causes plywood sheathing to delaminate. Waferboard roof sheathing will begin to crumble and may also turn black from excess moisture.

The results of poor attic ventilation can be very serious and costly to repair. Moisture buildup in the attic can cause rusting of nails and gussets in trusses, delamination of the roof sheathing, and rotting of the wood structural members. Repairing the situation can be very costly and should be classified as a major repair.


## Cathedral and Flat Roofs

Ventilation in cathedral ceilings and flat roofs is a special concern, and they're often inadequately ventilated. It is not unusual for the home inspector to find the roof over a cathedral ceiling to be weakened by delaminated sheathing due to poor ventilation. Special care should always be taken to test each step during the roof inspection over a cathedral ceiling.

As noted on page 77, there must be adequate ventilation space between the insulation and the roof sheathing in the cathedral ceiling (see drawing on that page). Cathedral ceilings should have both soffits vents and a ridge vent to move the air through this space. There should also be vents at both ends of each rafter bay.

*Personal Note*

*"I did an inspection where the homeowner had a misting system in his attic to tend to his marijuana plants. The moisture may have been healthy for the plants, but not the attic. No ventilation system on earth could have stopped the delamination and wood rot present in that attic. The word stupid comes to mind, doesn't it?"*

*Roy Newcomer*

*"One of my inspectors put his foot through a cathedral ceiling. He'd been careful when first getting on the roof and had tested several steps. Finding no signs of a weakened condition, he decided the roof was all right and he got overconfident. His next vigorous step plunged him right through the ceiling.*

*"Don't assume that the roof is in the same condition throughout. Small spots of deteriorating sheathing may be present."*

*Roy Newcomer*

A flat roof cavity may not be insulated, and *many were not intended to be*. When insulation is added, the problem of proper ventilation arises. One approach is the use of continuous soffit vents at both edges of the roof as shown in the drawing of the flat roof below.

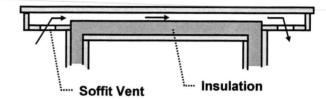

··· **Soffit Vent**       ···· **Insulation**

Another approach is to install wooden strapping members over the roof joists to create a deeper space under the roof. The insulation rests between the roof joists, and the roof sheathing is then installed on top of the strapping. This leaves a ventilating space between the two. In some cases, a vapor barrier and rigid board insulation is installed *over* the roof sheathing in an attempt to insulate the roof. No ventilating space is present.

Whatever the means of ventilating cathedral ceilings and insulated flat roofs, these areas remain susceptible to moisture buildup. Even if problems are not present, customers should be cautioned to watch them carefully for any new problems that may arise.

## Inspecting the Ventilation

The ventilation inspection takes places primarily on the roof, in the attic, and in the crawl space. The home inspector must do the following during the inspection:

- Identify the type and location of vents.
- Report on the condition of vents.
- Judge the adequacy of ventilation.
- Check power ventilators to see if they operate.

The home inspector should watch for the following conditions during the inspection of ventilation in the home:

- **Evidence of poor ventilation:** Reading the signs of poor ventilation is the first consideration during the inspection. In the attic, delaminated roof sheathing is a sign as is rust on roofing nails and on truss gusset plates. Rotting structural members may be a sign of poor ventilation, although rot may be present from a localized roof leak. Photo #31 shows **delaminated roof sheathing and rotting rafters** as the result of poor ventilation.

Frost on roofing nails in winter indicate condensation of moisture in the attic air, also a sign of poor ventilation. Feel the insulation. Consistently damp or wet insulation means that the insulation isn't being properly ventilated.

In the crawl space, delamination of the plywood subflooring indicates poor ventilation. Another sign can be moisture laden insulation falling from the floor over the crawl space.

- **Inadequate ventilation:** The home inspector should confirm the presence of adequate free vent area. This can be done by a rough estimate. Remember the attic requirements — **1 square foot of free vent for every 300 square feet of attic floor**. For example, if the attic is roughly 30' by 40' or 1200 square feet, there should be 4 square feet of free vent (1200 ÷ 300) or 8 square feet if louvers are present. You can do similar calculations for the crawl space, using the requirements stated earlier.

Another judgment of ventilation adequacy is the number and location of vents. Remember, there should be split vents in the attic and perimeter vents on opposite walls in the crawl space. Of course, adequacy can be judged by the condition of the structural members in the attic and crawl space. If vents are too few or they're too small, advise customers of what is required to improve ventilation.

- **Blocked, covered, or damaged vents:** Examine the condition of the vents. Make sure that vents are not blocked with insulation, especially soffit vents. Photo #32 shows a **blocked soffit vent**. Here, the homeowner has stuffed insulation over the soffit baffle, thinking he was insulating better. A so-called better insulation job should be sacrificed for good ventilation. We advised the customer to have the insulation removed from the vent area. Take another look at **Photos #21 and #26** for other examples of blocked soffit vents.

In general, vents should be in good condition and not obstructed with debris or animal nests. Dirty screening can

---

**INSPECTING VENTILATION**

- Evidence of poor ventilation
- Inadequate ventilation
- Blocked, covered, or damaged vents
- Inoperable ventilator
- Improper exhaust venting

---

↖ *Photo #31*

← *Photo #32*

cut down the operation of the vent. Point out any broken or missing louvers. Watch for gable vents that are covered with plastic or wood in an effort to save heat. Keep an eye out in summer when these covers may be set aside and advise customers to throw them out and not use them during the winter.

- **Inoperable power ventilator:** Try the power ventilator to see if it's working. Also check to see if a gable-mounted power ventilator is cutting down on the amount of free vent area required in the attic.

- **Improper exhaust venting:** Locate kitchen and bathroom exhaust vents and the plumbing vent stack in the attic area to be sure they exhaust through the roof. Also pay attention to their condition — holes in the vent piping and open joints will throw moist air into the attic. Roof leaks onto plumbing vents can cause holes from corrosion.

## Reporting Your Findings

Reporting on ventilation depends a great deal on how your inspection report is laid out. Roof vents may be reported on your roof page, for example, while attic conditions are on the attic page. In any case, be sure to cover these basic areas in your reporting:

- **Roof vents:** You'll be noting the type of attic and roof vents during the inspection of the roof and exterior of the home. First, identify the type of vents present such as soffit vents, ridge venting, roof-top vents, gable vents, turbine, and power ventilators. Indicate as many different kinds as you've found. Note their condition and record specific defects such as broken louvers in gable vents, broken roof-top vents, vents covered by insulation or boarded up for the season, and so on.

- **Attic:** We suggest that you have a special area for recording evidence of **moisture or condensation** in the attic. By the way, it's important to write whether the attic is finished and you couldn't see behind the construction to determine if condensation is present. Make use of the term *not visible* in this situation. Record the results of poor ventilation, noting any deterioration in the sheathing and roof framing.

Note whether or not fans and plumbing vents are exhausted to the attic or through the roof. Record the condition of such items, noting defects such as holes and open joints in vent piping. Don't forget to record if a power ventilator is not operating.

- **Crawl space:** First, note whether perimeter vents are present or not. Then write your findings, including situations where vents are blocked or there isn't a sufficient amount of vents present.

---

**DON'T EVER MISS**

- Signs of inadequate ventilation
- Blocked soffit vents
- Exhaust venting terminating in attic

---

# WORKSHEET

*Test yourself on the following questions.*
*Answers appear on page 98.*

1. According to most standards of practice, the home inspector is required to enter underfloor crawl spaces and attic spaces when:

   A. The home inspector can walk upright in the spaces.
   B. Access is easy.
   C. Access is not obstructed, won't damage property, or be dangerous.
   D. The inspector doesn't have to enter crawl spaces or attics.

2. If a crawl space has a vapor barrier, what amount of free vent space is required?

   A. 1 square foot for every 150 square feet of floor area
   B. 1 square foot for every 300 square feet of floor area
   C. 1 square foot for every 500 square feet of floor area
   D. 1 square foot for every 1500 square feet of floor area

3. Which of the following would not be considered adequate venting for a crawl space?

   A. A single perimeter vent on the north wall
   B. Perimeter vents on the south and north walls
   C. Perimeter vents on the east and west walls
   D. Perimeter vents on all four walls

4. What method of attic ventilation is considered to be the most effective?

   A. A gable vent at each gable
   B. A combination of soffit and ridge vents
   C. Soffit vents at each eave
   D. A ridge vent along the full length of the roof

5. Identify the types of vents shown in the drawings below.

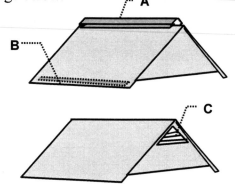

6. Which of the following vents is normally controlled by a thermostat?

   A. Turbine vent
   B. Roof-top vent
   C. Ridge vent
   D. Power ventilator

7. Which of the following conditions would not be a sign of poor ventilation in an attic?

   A. Cracked rafters
   B. Delaminated sheathing
   C. Rusted gusset plates on roof trusses
   D. Frost on roofing nails

8. Which of the following conditions would be a sign of poor ventilation in a crawl space?

   A. Cracked floor joists
   B. Rotted subfloor
   C. Missing vapor barrier
   D. Special radon reduction devices present

9. How would you calculate the amount of ventilation needed in a 1000 square foot attic?

   A. Divide 1000 by 150.
   B. Divide 1000 by 300.
   C. Divide 1000 by 500.
   D. Divide 1000 by 1500.

# Chapter Eleven
# INSPECTING THE ATTIC

We've presented most of what you need to know about attic inspection in bits and pieces throughout this guide. This chapter will put the attic inspection all together for you.

The following items are included in the attic inspection:

- The entryway
- Structural members
- Evidence of water penetration and condensation
- Insulation and vapor barrier
- Vents and ventilation
- Chimney chase
- Electrical safety
- Exhaust venting
- Condition of equipment such as air handler, fans, and so on

Always report the means of access to the attic, whether it's the attic stairs, pull-down stairs, or through a hatch in the ceiling. And always report whether attic access is limited in any way. Some attic spaces may be too small to actually get into and move around in, and the attic inspection may have to be done by standing on a ladder in the hall. Whatever the case, be sure to accurately describe the **methods used** to observe the attic. Let your customers know that existing defects may not be found when access is limited. This is for your own protection.

Make every attempt to get down the entire length of the attic during the inspection, but be careful where you walk. Some attics have plank floors, some a plank walkway down the center, and others only exposed ceiling joists. Whatever you do, don't step between the ceiling joists. Your foot will go right through the ceiling. Watch out too when walking on planks that aren't nailed down. Lighting is likely to be insufficient in the attic, so take your flashlight with you. And, as mentioned earlier, if you are susceptible to respiratory aliments, you might want to wear a dust mask to protect yourself from insulation fibers.

The order of attic inspection is up to the home inspector. It's wise to develop a routine so that nothing is missed. Some home inspectors will make one slow trip up and down the attic, examining all aspects at the same time. Others may examine the

*Guide Note*
*Pages 97 to 103 present a summary of the attic inspection.*

*For Beginning Inspectors*
*Perform a complete inspection of your own (or a friend's) attic, paying attention to the order in which you do it. Try to develop a routine that helps you not to miss anything.*

roof structure first, then concentrate on insulation, ventilation, the chimney, and so on in turn.

---

**ATTIC ACCESS**

- Formal stairway

- Pull-down stairs

- Scuttle

---

## The Entryway

Start the attic inspection by examining the entryway to the attic. **Formal stairs** may be present with a door at the bottom. Examine them as you would any other staircase for the condition of risers and treads, secure handrails, and proper lighting. Stairs that open directly into the attic space should have a secure guardrail around the stairwell. When the stairway opens into the attic in this way, it should be considered part of the attic space requiring insulation. Heat from the house should not be allowed to enter into the stairwell. Try to determine if the door at the bottom closes tightly and if there is insulation present under the stairs and along the stairway walls. (For stairway inspection, see pages 35 to 36.)

Be careful when you open the **pull-down stairway**. Sometimes, cables are broken or the counterweights are detached, and the stairs may come down too quickly. Examine the stairway for loose treads, broken stringers, and loose or missing bolts before you climb up. If there are no provisions to insulate this type of stairway such as a cover or insulated box over the opening, suggest to customers that they might want to add insulation here. Even a sheet of plywood over the opening would cut down on warm air entering the attic.

The entry to the attic may be a simple **scuttle or hatch** in a hall ceiling or closet, and you'll probably need a ladder to get in. Note if the hatch cover fits securely to prevent warm air from entering the attic. Again, if insulation isn't present on the attic side of the hatch cover, you can suggest to customers that insulation be added.

Once you gain access to the attic, take a moment to examine the area before you go charging in. Determine if the lighting is adequate, if conditions are safe, whether you'll be able to access the entire space or not, and so on. Check the floor so you'll know how you're going to get from one end to the other. And make a note of what's in the attic such as chimneys, exhaust venting, or air handlers that you don't want to miss. It pays to get yourself organized before you begin the inspection.

*Worksheet Answers (page 96)*
1. *C*
2. *D*
3. *A*
4. *B*
5. *A is a ridge vent.*
   *B is a soffit vent.*
   *C is a gable vent.*
6. *D*
7. *A*
8. *B*
9. *B*

## Structure Review

During the attic inspection, the home inspector will inspect the **roof structure** of the home, including the following structural members:

- Rafters, collar ties, and knee walls
- Trusses
- Roof sheathing
- Ceiling structure

Check down the rafters for any cracks, warping, and sagging. Be sure rafters are securely fastened to all other framing members — ridge, plate, and joist connections and supporting structures. Rafters should be reinforced if they're supporting heavy equipment on the roof and should not be cut for any reason. Watch for **rafter spread**, a condition where the roof load bearing on the rafters forces them outwards. Check the ridge board for cracking and twisting. If collar ties are present, inspect them for buckling and secure nailing to the rafters.

If **roof trusses** are present, watch for any cracked, cut, or missing truss members and check gussets for secure attachment. Be sure that lateral bracing is present along the diagonal webs of the truss if it's required. (Usually, there's a tag on the truss that says this.) Another condition that can be present is **truss uplift**, a phenomenon where the bottom chord of a roof truss bows upward during the cold months and returns to normal position during the warmer months. Often, it will carry the ceiling with it, and you can often see cracks at the ceiling-wall intersection in the rooms below.

Inspect the roof sheathing and check for sags, warping, and buckling of the panels. Check that nails and clips are in place and holding. Report any indication of delamination of the sheathing. Photo #33 shows **plywood delamination**. Take another look at **Photo #31**, which shows plywood delamination and deterioration of the rafters against the sheathing. These cases are the result of poor ventilation in the attic.

As you inspect structural members, keep an eye out for any signs of water penetration, condensation, or deterioration in any of the members. Photo #34 shows **leakage into the attic**. Here, you can see water stains along the roof edge and along the chimney chase where water has been leaking in. Wood should

---

### ATTIC STRUCTURE

- Cracked, bowing, or twisting rafters
- Rafter spread and sidewall separation
- Sags in the roof
- Delaminated plywood
- Loose truss fastenings, bowing trusses, cracked, cut, or missing members
- Deterioration of structural members
- Water penetration

← *Photo #33*

← *Photo #34*

be probed in these areas for wood rot. The customer should be advised to have someone locate and repair the source of the leaks.

## Water Penetration

The home inspector should examine the attic for any evidence of water leaking into the house. Examine the sheathing and rafters for any signs of deterioration from water penetrating through the roof surface. Inspect the chimney to see if there is any deterioration in the mortar and bricks from leaking flashings. Check carefully around other roof openings such as vents and piping to be sure they are water tight.

*Photo #35* →

Photo #35 shows a **swimming pool in the attic**. When we first saw this children's pool, we thought it was merely being stored in the attic. However, a closer look showed that it had been cut to fit neatly between the floor joists and slid into place to catch water from a roof leak. Obviously, that leak had been going on for a long time. Instead of fixing the leak, the homeowner had devised this solution.

*Personal Note*

*"I found the children's swimming pool in the attic during one of my inspections. When I came down from the attic, I told my customers I had good news and bad news. I told them they had an indoor pool, but they'd have to use it in the attic. Otherwise, the roof leak would wreck the ceiling.*

*"Be sure to investigate all of the attic and everything in it. I could have so easily dismissed this pool, thinking it was just being stored up there. Be suspicious about buckets and basins sitting in the attic."*

*Roy Newcomer*

## Insulation

Inspect the attic insulation as instructed in pages 68 to 84. First, identify the type of insulation, then estimate the average inches. Check for proper insulation and look for gaps, blocked vents, and so on. Feel the insulation for dampness and determine the condition of the insulation. Try to locate a vapor barrier against the warm side of the house. If plastic rigid insulation is present and exposed, report it as a **safety hazard**. Advise customers to cover the rigid insulation with drywall for fire safety.

*Photo #36* →

## Ventilation

Inspect the attic ventilation as instructed on pages 86 to 95. Locate and identify the type of vents. Check the vents for condition and blockage. Watch for any vents that are covered over completely in the winter. Photo #36 shows a **covered gable vent** at the far end of the attic. Customers should be advised that the vent should remain uncovered year-round. **Photos #21 and #32** show blocked soffit vents. Determine the adequacy of ventilation based on the amount of venting present and any signs of high moisture content in the attic and signs of condensation or deterioration on framing members. **Photos #26, #31, and #33** show the results of poor ventilation. Take the time to take another look.

Watch for improper venting of plumbing and house exhaust fans. Photo #37 shows **plumbing ve nts** in the attic area. Here, the vents are properly exhausted through the roof, and the waferboard sheathing is in good shape with no indication of flashing leaks around the vent penetrations.

## The Chimney

Check the chimney chase in the attic for condition and signs of leaking at the roof penetration. In a **masonry chimney**, check the mortar and bricks for condition. If the chimney is not lined (has no separate flue), any cracked or missing bricks or missing mortar means that carbon monoxide gases are escaping directly into the attic. This should be reported as a **safety hazard**. Photo #38 shows a **deteriorating chimney**. Notice the crack that runs the height of this old chimney, the missing brick, and missing mortar in places. This unlined chimney was not in good condition above the roof either. We reported it as a safety hazard.

**Metal chimneys** passing through the attic must be isolated from the floor and roof framing with approved fittings. Metal chimneys must have a 2" clearance from combustibles, and the home inspector may find this clearance open in the attic floor. This is a **fire hazard**, as a fire starting in the basement can move right up the chimney into the attic. An appropriate fitting around this clearance opening should be a non-combustible material such as sheet metal. Inspect a metal chimney for corrosion, rusting, and open or loose joints. Look for soot or creosote buildup around the joints. Any discoloration indicates that smoke and carbon monoxide gases are leaking into the attic. Report any open or leaking joints as a **safety hazard**.

---

### ATTIC INSULATION AND VENTILATION

- Wet, damaged, or improperly installed insulation
- Missing vapor barriers
- Exposed plastic insulation
- Suspected asbestos
- Signs of inadequate ventilation
- Blocked soffit vents
- Exhaust venting terminating in the attic

↖ *Photos #37 and #38*

*Guide Note*
*For more information on chimney inspection, see another of our guides — A Home Inspector's Guide to Inspecting Roofs.*

*Study Note*

*For more information on inspecting the electrical system, see another of our guides — A Home Inspector's Guide to Inspecting Electrical.*

*Personal Note*

*"One of my inspectors found a situation of immediate danger when he saw knob-and-tube wiring sparking onto old cellulose insulation in an attic. He told the homeowner and the REALTOR on the spot, suggesting they call an electrician immediately. He was a hero, although the homeowner chose not to listen to his warnings. The home burned down a week later!"*

*Roy Newcomer*

## Electrical Safety

The home inspector's main concern with the electrical wiring in the attic is whether it poses any safety hazards.

Inspect the visible wiring, junction boxes, and light fixtures in the attic carefully. The home inspector should watch out for and report the following **safety hazards**:

- **Uncovered or missing junction boxes:** We've seen uncovered junction boxes half buried in the insulation and even splices with no junction boxes. These practices are terribly unsafe and must be reported and remedied.

- **Brittle and damaged knob-and-tube wiring:** Knob-and-tube wiring is old wiring which can become damaged or brittle after years of overheating. Check out the wiring and the connections for condition. This wiring should *never be buried under the insulation* because of the likelihood of its being damaged in some way. It's a definite safety hazard.

When knob-and-tube wiring is present, always be sure to check out any light fixtures present, where connections are likely to be poor. We always suggest that these old fixtures with cloth insulated wire be replaced for safety's sake.

- **Extension-cord wiring:** The NEC forbids the use of extension cords as part of the permanent wiring in a home. This is especially dangerous in the attic, where extension cords are taking the place of inadequate wiring to light the attic. Report the use of extension cords as a safety hazard.

- **Recessed lights covered with insulation:** Unless a recessed ceiling light fixture is rated for an insulated ceiling, there should be a 3" clearance between the fixture and the insulation. It's impossible to tell if the light is rated or not. The home inspector is not required to dig out the light to check, but should advise the customer to do so and provide baffling if the light is not rated. The inspector should report the light as a safety hazard if it's covered with insulation.

You'll be a hero if you find safety hazards in the attic. Some homeowners never go into the attic to look, and dangerous situations can be sitting there waiting for something to happen.

## Equipment in the Attic

The home inspector will inspect fans and power ventilators and any heating and cooling equipment located in the attic. If an air handler is present, care should be taken to inspect the condition of the condensate tray and auxiliary drip pan for any signs of corrosion, rusting, or leaking into the attic.

## Reporting Your Findings

When you report on your inspection of the attic, always begin by noting how you accessed the attic. Be specific. Write whether you inspected it from the scuttle, the pull down, the stairs, whether you got into the attic and viewed all of it or were not able to view it at all. This is for your own protection. There can be serious defects present under the roof that you're not able to see. Be sure that your report clearly states this.

We've already discussed reporting on the insulation and ventilation inspection of the attic. Here's what is left to report on:

- **Roof structure:** Identify structure as rafters or trusses and note their condition in your report. Then identify the sheathing material and its condition. Make a note of any defects such as loose gusset plates on trusses, cracked rafters, and other structural problems you have observed.

- **Chimney:** Report on the chimney chase as viewed from the attic, noting missing bricks or mortar as a safety hazard.

- **Water penetration:** Don't ever miss recording roof leaks. Make a point of distinguishing between old and fresh water stains if you wish, but don't overlook writing about water stains just because they're old. You probably can't determine whether the roof is still leaking or not.

- **Electrical:** Always record your findings of electrical safety hazards as indicated on page 102. If you've reported these safety hazards on the attic page, it's a good idea to repeat them on a summary page of your report.

---

**DON'T EVER MISS**

- Deteriorating structural members
- Water penetration
- Improperly installed insulation
- Signs of inadequate ventilation
- Deteriorating chimney
- Electrical safety hazards

---

# EXAM

*A Home Inspector's Guide to Inspecting Interiors, Insulation, Ventilation* has presented quite a few details. Now you have an opportunity to test yourself to see how well you've learned them.

*Requirements to receive 7 Continuing Education Units (CEUs):*
To complete the following exam, fill out the *answer sheet* (at the back of the exam), and mail it, along with a $35 check, to:

 American Home Inspectors Training Institute
 212 Wisconsin Avenue
 Waukesha, WI  53186
*Please advise as to which organization you are seeking CEUs.*

It will be necessary to pass the exam with at least a 75% passing grade in order to receive CEUs. If you do not pass the exam, a new exam will be mailed to you so that you can try again.

*Roy Newcomer*

Name:_____

Address_____

_____  Phone:_____

*Circle the letter for the correct answer for each of the following questions.*

1. Which action is required by most standards of practice during the interior inspection?

   A. Required to report signs of water penetration into the building or signs of abnormal condensation
   B. Required to observe draperies, blinds, or other window treatments
   C. Required to inspect recreational facilities
   D. Required to observe paint, wallpaper, and other finish treatments on the interior walls, ceilings, and floors

2. Which of the following actions is required by during the insulation, ventilation, and attic inspections?

   A. Required to report on concealed insulation and vapor retarders.
   B. Required to report on venting equipment which is integral with household appliances.
   C. Required to enter the attic space when entry could damage property.
   D. Required to observe insulation and vapor retarders in unfinished spaces.

3. Which is the most recent method of applying plaster to walls and ceilings?

   A. 1 or 2 layers over a gypsum board lath
   B. 3 layers over 1" by 4" lath
   C. 3 layers over a wire mesh
   D. 1 layer over plywood

4. What repair would not be acceptable with powdery plaster?

   A. Remove the powdery plaster and replaster the area.
   B. Cover the powdery plaster with a finish plywood.
   C. Wet the plaster to reharden it.
   D. Cover the powdery plaster with drywall.

5. What sort of crack in a plaster ceiling can be caused by truss uplift?

   A. A crack running alongside a girder the length of the house
   B. A crack along the junction of the ceiling and the walls
   C. An angled crack from corner to corner of the ceiling
   D. A network of cracks over the ceiling

6. What condition would not be the cause of a sag in a plaster ceiling?

   A. Detachment of the plaster from the lath
   B. Contraction and expansion of a lintel
   C. Detachment of the lath from the ceiling joists
   D. Water damage in the ceiling

7. Which of the following is a true statement?

   A. Drywall sheets cannot crack from structural stresses.
   B. Drywall does not absorb moisture.
   C. Ceiling drywall can sag if it is less than 1/2" thick.
   D. Drywall is made of a material entirely different from plaster.

8. What is the ceiling covering shown in Photo #1 at the back of this guide?

   A. Texturing
   B. Painted drywall
   C. Decorative plywood
   D. Acoustic tiles

9. What should the home inspector do if he or she finds a ceiling as shown in Photo #20? *Circle 3 answers.*

   A. Report it as water damaged in the inspection report.
   B. Recommend that new drywall be installed to cover the stain.
   C. Point out the area of concern to the customer.
   D. Investigate further to find the cause of the stain.

10. What problem can be experienced when a softwood plank subfloor is exposed and used as the finish flooring?

    A. Sagging because of the thinness of the softwood
    B. Not being able to sand the floor smooth
    C. Trip hazards because of open joints
    D. Keeping the joints sealed against spills

11. What can be the cause of cracked natural tiles such as slate or stone?

    A. Installing the tiles over concrete
    B. A flooring system that's too rigid
    C. A flooring system that's too flexible
    D. Defective adhesive

12. In wood frame construction, what might be the cause of a floor sloping toward an interior wall?

    A. Interior walls shrinking more than outer walls and pulling floor down
    B. Foundation settlement pulling the floor lower at the center of the house

13. In slab-on-grade construction, what is a sign that the slab has shrunk and pulled away from the foundation?

   A. Open spaces between the slab and interior walls
   B. A floor sloping toward an interior wall
   C. Cracking along the junction of the floor and outer walls
   D. A floor sloping toward an exterior wall

14. What is the cause of floor squeaks?

   A. Subfloor rubbing against the joists
   B. Stiffness in the floor joists
   C. Defective adhesive under resilient floor coverings
   D. Nails moving in and out of the joists

15. Which of the following is an example of safety glazing in a window?

   A. Double lights
   B. Tempered glass
   C. Thermal-pane
   D. Translucent glass

16. The home inspector is required to operate a representative number of windows.

   A. True
   B. False

17. What condition would not be a problem with self-storing windows?

   A. Missing storm or screen
   B. Broken crank
   C. Blocked drain holes
   D. Rotted sill

18. The home inspector should observe a representative number of windows in the home.

   A. True
   B. False

19. What condition does discoloration or cloudiness between the lights of a thermal-pane window indicate?

   A. Framing out of square
   B. Rotted sill
   C. Casing pulling away from the wall
   D. A leaking seal

20. What type of door has a honeycomb cardboard interior set in wood rails and jambs?

   A. Panel door
   B. Louvered door
   C. Hollow core door
   D. French door

21. In what location in the home would an interior hollow core door not be appropriate?

   A. As a bathroom door
   B. As the door between the living area and the garage

22. Which of the following stairway conditions should be reported as a safety hazard? *Circle 4 answers.*

   A. The presence of winders
   B. Missing handrail
   C. Uneven or loose risers or treads
   D. Balusters at 4" apart
   E. A flat landing at a turn in the stairs
   F. A window at the bottom of the stairs that is 42" off the floor
   G. Loose handrail
   H. Balusters at 6" apart

23. In cantilevered balcony construction, what structural members provide support?

   A. A back wall
   B. A girder between 2 bearing walls
   C. Chains from the ceiling joists
   D. Floor joists

24. What is the requirement regarding balcony balusters?

    *A.* They should be metal.
    *B.* They should be a maximum of 4" apart.
    *C.* They should be a maximum of 8" apart.
    *D.* Balusters are not allowed on balconies.

25. What is a zero clearance fireplace?

    *A.* An old fireplace with a small firebox and decorative tile border
    *B.* A prefabricated wall-mounted or freestanding metal fireplace
    *C.* An opening with a connection to the chimney intended for fireplace installation at a later date
    *D.* A conventional masonry fireplace

26. The home inspector is <u>not</u> required to:

    *A.* Light or put out fireplace fires.
    *B.* Inspect the fireplace damper.
    *C.* Observe fireplace clearance from combustibles.
    *D.* Inspect a masonry fireplace with a metal firebox.

27. Which of the following conditions might cause a fireplace to smoke? *Circle 2 answers.*

    *A.* Negative pressure condition in the home
    *B.* Soot and creosote buildup in the flue
    *C.* A poorly supported hearth
    *D.* Using outside air for combustion

28. According to most standards, when should a wood stove be inspected?

    *A.* Never
    *B.* Only if it's UL rated
    *C.* If it's the primary heating source

29. When should a wood stove sit on a protective non-combustible pad?

    *A.* Over a concrete floor
    *B.* Over a wood floor

30. Where are GFCI's recommended in a home?

    *A.* In long hallways
    *B.* In all rooms
    *C.* On exterior walls
    *D.* At outlets within 6' of water

31. During the room-by-room inspection, the home inspector is required to:

    *A.* Test every electrical outlet.
    *B.* Flush every toilet.
    *C.* Open every window.
    *D.* Fill up the bathtub.

32. Which of the following photos found in this guide shows an illegal plumbing practice?

    *A.* Photo #10
    *B.* Photo #11

33. Which of the following shower findings must be reported and never missed?

    *A.* One with a tub and shower combination
    *B.* The use of ceramic tiles as a shower surround
    *C.* The presence of a metal shower pan
    *D.* One with a fiberglass surround

34. What ought to be the home inspector's number one concern when inspecting the bathroom?

    *A.* Leaks, water damage, and wood rot
    *B.* The date of manufacture stamped under the toilet tank lid
    *C.* Malfunctioning faucets
    *D.* The noise level of the exhaust fan

35. In general, which type of insulation has the lowest R-value?

    A. Batts and blankets
    B. Loose fill insulation
    C. Rigid board insulation
    D. Site-foamed insulation

36. If an insulation's R-value is 3.1/inch and the insulation is 6" thick, what is its R-rating?

    A. About R-3
    B. About R-9
    C. About R-12
    D. About R-19

37. What is kraft paper?

    A. A type of cellulose insulation
    B. A water-resistant asphalt impregnated paper used as a vapor barrier
    C. Reflective metal foil used as a facing on batts and blankets
    D. A batt insulation made from asbestos

38. Why was UFFI insulation considered to be a health hazard?

    A. It contains asbestos.
    B. Its glass fibers can be a skin irritant.
    C. It releases formaldehyde into the air.
    D. The plastic it contains can give off toxic fumes when it burns.

39. How can urethane foam insulation be recognized?

    A. By its yellowish orange color and shiny finish
    B. By its soft foamy appearance and light color
    C. By its concrete-like appearance
    D. By its pink color

40. Which of the following photos in the guide shows the presence of suspected asbestos insulation?

    A. Photo #21
    B. Photo #23
    C. Photo #26
    D. Photo #27

41. What condition should be reported in Photo #25?

    A. Insulation likely contains asbestos
    B. Presence of UFFI insulation
    C. Insulation installed upside-down
    D. Vapor barrier missing

42. Where would insulation not have to be installed?

    A. On floors over unheated areas
    B. On ceilings below unheated areas
    C. Behind knee walls in a finished attic
    D. On pipes and ducts in heated areas

43. Which of the following statements is true?

    A. A cathedral roof does not need to be insulated.
    B. The vapor barrier in the ceiling under the attic should be on the ceiling side of the insulation.
    C. The vapor barrier in the floor over the crawl space should be on the crawl space side of the insulation.
    D. A flat roof is always insulated.

44. What R-rating is recommended for floors over unheated spaces?

    A. R-13
    B. R-19
    C. R-30 or R-38

45. What is the true purpose of the ventilation process?

    A. To reduce the home's heat loss
    B. To reduce the home's heat load
    C. To remove moist air from the home
    D. To stop radon from entering the home

46. What amount of free vent space is required for an attic?

    A. 1 square foot for every 150 square feet
    B. 1 square foot for every 300 square feet

47. When should attic vents be kept open?

    A. In summer only
    B. In winter only
    C. In the spring and fall only
    D. At all times

48. What is the source of radon in a home?

    A. It's released from UFFI insulation in the home.
    B. It comes from crumbling plaster insulation on boiler pipes.
    C. It comes from the breakdown of uranium in the soil and rocks beneath the home.
    D. It's released from urethane site-foamed insulation when it burns.

49. What type of roof is often inadequately ventilated?

    A. One with a ridge vent
    B. One over a cathedral ceiling
    C. One with roof-top and gable vents
    D. One over an unheated attic

50. What condition can be a sign of condensation in the attic during the winter?

    A. Presence of an ice dam at the roof edge
    B. Frost on the roof surface
    C. Frost on the roofing nails in the attic
    D. All of the above

51. What is the cause of the delaminated sheathing as shown in Photo #21?

    A. Insulation blocking soffit vents
    B. An inadequate amount of insulation
    C. Insulation installed upside-down
    D. Absence of a vapor barrier

52. What condition in the attic should be reported as the source of a ventilation problem?

    A. Louvered gable vents
    B. Presence of a whole house fan
    C. Open joints in exhaust vents in the attic
    D. Split ventilation

53. What condition is a sign of too much moisture in the attic?

    A. Rafter spread
    B. Delaminated plywood sheathing
    C. Loose gussets on trusses
    D. A cracked ridge board

54. Which conditions found during the attic inspection should be reported as safety hazards? *Circle 6 answers.*

    A. Knob-and-tube wiring buried under the insulation
    B. Plywood delamination
    C. Open joints in metal chimney
    D. Plumbing vents passing through attic
    E. Covered gable vent
    F. Uncovered junction box
    G. A roof leak
    H. Exposed polystyrene insulation
    I. Missing bricks in masonry chimney
    J. Water stains
    K. Buckled collar ties
    L. Fiberglass loose fill insulation
    M. Extension-cord wiring

55. What should the home inspector do if a recessed ceiling light is buried under the insulation?

    A. Dig out the light and check to see if it's rated for an insulated ceiling.
    B. Advise the customer a safety hazard may exist if the light is not rated to be covered.
    C. Ignore the light and don't report anything.
    D. Suggest the light be removed.

# GLOSSARY

**Acoustic tiles**  A ceiling tile made of fiber board, fiberglass, cork, or mineral particles.

**Airkrete**  A site-foamed insulation made of a mixture of cement containing syrup and air.

**Asbestos**  A mineral fiber found in rocks, used in home products including insulation.  The EPA considers asbestos fibers to be a health hazard.

**Awning window**  A window hinged at the top to open outward.

**Balcony**  A platform protruding into the living space with no visible connection to the floor below.

**Balusters**  The vertical poles that support the railing of a staircase.

**Baseboard**  A trim piece used at the intersection of the wall and floor.

**Batts**  Pre-cut lengths of fibrous insulation manufactured to fit between studs and joists, made of fiberglass or rock wool.

**Bead board**  Rigid board polystyrene insulation produced by expanding and fusing granular pellets.

**Bifold door**  A 2-section door hinged in the middle to allow one section to fold back on the other before the door is swung to the side.

**Blankets**  Continuous rolls of fibrous insulation manufactured to fit between studs and joists, made of fiberglass or rock wool.

**Cantilever**  An extension of the floor structure which depends on the strength of the unsupported portion of the girder or joists to carry the load of the structure.

**Casement window**  A window hinged at the side to open outward.

**Casing**  The trim pieces used around window and door frames where they meet the wall.

**Cellulose insulation**  An insulating material made of shredded recycled newspaper or wood fibers treated with a fire retardent, used as loose fill.

**Ceramic tiles**  Hard fired clay tiles that can be glazed or unglazed, used as a floor covering.

**Cornice molding**  A trim piece used at the intersection of the wall and ceiling.

**Creosote**  A byproduct of a wood burning fire.

**Damper**  In a fireplace, a metal plate that closes the throat when the fireplace is not in use.

**Decorative fireplace**  A false fireplace designed to imitate the look of a real fireplace.

**Delamination**  A deterioration process during which the layers in the laminated plywood panel begin to separate.

**Double hung window**  A window with two sashes which both move.

**Drywall**  A premanufactured plaster sheet covered with paper, used as a wall and ceiling facing.

**Extruded polystyrene insulation**  Rigid board insulation produced by extrusion.

**Facing**  The material applied directly to studs and joists to form walls and ceilings.

**Fiberglass insulation**  An insulating material made of threads of glass covered with a coating that binds the fibers in place, used as batts, blankets, loose fill, and rigid board.

**Finish flooring**  The flooring applied over the subfloor that provides the finish floor for the home.

**Firebox**  In a fireplace, the open chamber in which the fire burns.

**Firebrick**  A special brick designed to withstand high temperatures.

**Fireplace insert**  A metal stove with a door that is totally or partially inserted into a fireplace.

**Fixed-pane window**  A window that does not open or close.

**Flue**  The channel that carries smoke and gasses from a fireplace out of the home.

**Flue liner**  A material that lines the chimney, usually terra cotta tiles.

**Flush door**  A door made of veneer glued to a solid or hollow core.

**Functional drainage**  A determination of whether water drains fast enough and completely.

**Functional flow**  A determination of whether water flows with enough pressure and volume.

**Gable vent**  A vent located at the gable ends of the roof.

**GFCI**  An abbreviation for ground fault circuit interrupter, a monitoring device that will trip after a ground fault is detected, stopping the flow of electricity to a circuit.

**GFCI tester**  A testing device used to test the operation of GFCI outlets and 3-slot outlets for polarity and grounding.

**Glazing**  The window pane made of glass or other material.

**Gypsum**  A common mineral also called sulfate of calcium, used in plaster.

**Hardwood**  Woods like oak, beech, birch, hard pine, maple, pecan, and walnut.

**Head**  The top piece in a window frame.

**Hearth**  The floor of the fireplace, usually poured concrete about 4" thick.

**Hollow core door**  A door with a honeycomb or patterned cardboard core framed in solid rails and jambs and covered with a veneer.

**Hopper window**  A window hinged at the bottom to open inward.

**Jalousie window**  A window with narrow strips of glass that move together, lifting out from the bottom as the window opens.

**Jambs**  The side pieces in a window frame.

**Joint cement**  A plaster-like paste used to seal the joints between sheets of drywall.

**Joists**  Horizontal members of a floor system that carry the weight of the floor to the foundation, girders, or load-bearing walls.

**Junction box**  A covered metal box used to protect connections or junctions in an electrical circuit.

**Kalsomine**  A mixture of glue, pigment, and water that was once used as a finish on plaster ceilings.

**Knee wall**  Supporting wall running from the ceiling joists to rafters which prevents rafter sag.

**Knob-and-tube wiring** Old branch circuit wiring using ceramic knobs to secure wires to surfaces and tubes to pass wires through framing members.

**Kraft paper** A water-resistant asphalt impregnated paper that can be used as a vapor barrier.

**Laminated glass** A multiple light glazing with a sticky plastic inner light that holds broken pieces of glass together when the glass breaks.

**Lath** Strips of wood, wire mesh, or gypsum board attached to studs and joists to form a base for plaster to adhere to.

**Light** Each layer of glass making up a window pane.

**Loose fill insulation** A loose insulating material poured or blown into place between wall studs and attic joists, made of fiberglass, rock wool, cellulose, or vermiculite.

**Louvered door** A door with plastic, wood, or cloth slats in a frame.

**Millboard** An insulating material made of asbestos and gypsum.

**Multi-pane window** A window with small pieces of glass set into wood or lead muntins.

**Multiple lights** Glazing consisting of two or three layers of glass or other material. Commonly called thermal-panes or Thermopanes, a tradename.

**Muntins** A grid of cross pieces of wood or lead that hold small panes of glass in a multi-pane window.

**Nail pops** A condition where nails cause bulges in drywall when they back out of the studs.

**Neon bulb tester** A testing device used to test 2-slot and 3-slot outlets for polarity and grounding.

**Nosing** The extension of the tread over a riser on a staircase.

**Open ground** Where an outlet is not grounded.

**Panel door** A solid door where the rails and stiles are grooved in such a way to hold an inset panel.

**Parquet flooring** 6-inch squares made up of six 1-inch strips of wood, laid at right angles to each other to provide a finish floor.

**PCi/L** An abbreviation for picocuries of gas per liter of air, the way radon is measured.

**Perlite** An insulating material made of volcanic rock, used as loose fill.

**Plaster** A powder made of gypsum and other aggregates that forms a paste when wet and a durable surface when applied and dried. Used as a wall and ceiling facing.

**Plywood** A building material made of three or more layers of wood sheets joined with glue.

**Pocket door** A door designed to slide sideways for concealment in a wall.

**Polystyrene insulation** A plastic foam insulation manufactured as rigid boards.

**Power ventilator** An electrically powered ventilator located at the gable end of the roof or on the roof surface between rafters.

**P-trap** A p-shaped trap commonly used today below fixtures which is usually vented.

**Quarry tiles** Hard fired clay tiles that can be glazed or unglazed, used as a floor covering.

**Radon**  A gas that occurs naturally when uranium in the soil and rocks breaks down.  The EPA considers radon as a health hazard.

**Rafters**  Structural members of a pitched roof that support the roof covering and transmit roof loads to bearing walls and beams below.

**Rafter spread**  A condition where the roof load bearing on the rafters forces them outwards.

**Rail**  The top piece in a window sash.

**Resilient floor coverings**  A range of tiles and sheet goods laid as a finish floor.  Includes linoleum, rubber tiles, and various vinyl sheets and tiles.

**Reverse polarity**  Where the hot wire is wired to the large slot in an electrical outlet and the neutral wire is wired to the small slot, the opposite way of how it should be done.

**Ridge vent**  A vent located on the roof ridge, running the length of the ridge.

**Rigid board insulation**  Wide rigid boards of insulation used on foundations and walls, made of fiberglass, wood fiberboard, or foamed plastics.

**Riser**  The vertical portion of a step on a staircase.

**Rock wool**  An insulating fibrous material made by blowing steam through molten rock or slag, used as batts, blankets, or loose fill.

**Roughed-in fireplace**  An opening left with a connection to the chimney intended to provide the space for a fireplace to be installed later.

**R-rating**  The total heat resistance for a given thickness of insulation and equal to the insulation's R-value times its thickness.

**R-value**  A number that represents an insulation material's resistance to heat flow per inch of thickness.

**Safety glazing**  Glazing that is held in place or is otherwise harmless when it breaks.  May be tempered or laminated glass.

**Sash**  The framework in a window that holds the glass or other material.

**Self-storing window**  A double hung window with a storm and screen permanently mounted in a 3-channel metal or plastic frame.

**Separation walls**  Walls between the garage and living area which must be covered with drywall for fire resistance.  Also called firewalls.

**Sill**  The bottom piece in a window frame.

**Single hung window**  A window with two sashes, only one of which moves.

**Site-foamed insulation**  Insulation made of syrups and reactants mixed at the building site and foamed in place.

**Slider window**  A window with a sash that moves horizontally.

**Smoke chamber**  In a fireplace, the area between the damper and the flue that guides smoke toward the flue.

**Smoke shelf**  In a fireplace, the base of the offset between the damper and the flue that interferes with and interrupts chimney downdrafts.

**Soffit**  The horizontal underside of the eave.

**Soffit vent**  A vent located in the soffit along each side of the roof.

**Softwood**  Woods like pine, fir, or cedar.

**Stile**  The side piece in a window sash.

**Stippling**  A texturing process where a stipple finish is sprayed over drywall.

**S-trap**  An s-shaped unvented trap once used under plumbing fixtures.

**Stringer**  Long diagonal member that supports a staircase.

**Subflooring**  Horizontal sheets or planks that transfer the load of the home's furnishings and people to the floor joists.

**Surround**  The wall around the bathtub, made of ceramic tile or a premolded fiberglass.

**Tempered glass**  Glass that shatters into small, smooth edged cubes when it breaks.

**Terrazzo**  A mix of marble chips and concrete laid in squares bordered by lead beading and polished smooth, used as a floor covering.

**Texturing**  Adding sand or other agents to a conventional finish on a wall or ceiling. Also a design put in the finish coat of plaster done manually with a trowel.

**Thermal-panes**  See *Multiple lights.*

**Throat**  In a fireplace, the top of the firebox.

**Tongue and groove**  Boards with the top edge of each board wider than the lower. When nailed in place, the top edges fit snug with no gaps.

**Tread**  The horizontal portion of a step on a staircase.

**Truss**  An engineered, prefabricated framing member used in floor and roof construction.

**Truss uplift**  A phenomenon where the bottom chord of a truss bows upward during the cold months and returns to its normal position during the warmer months.

**Turbine vent**  A vent with air powered vanes on a central rotating spindle, located on the roof face.

**UFFI**  An abbreviation for urea formaldehyde foam insulation which is made of an insulating material in a syrup form and a reactant to create a site-foamed insulation.

**Urethane insulation**  A plastic foam insulation manufactured as rigid boards or foamed in place at the building site.

**Vapor barrier**  A type of waterproof sheeting used to prevent the passage of moisture through a surface.

**Vermiculite**  An insulating material made from heating and expanded mica, used as loose fill.

**Wall studs**  Vertical wall framing members.

**Whole house fan**  A fan with an intake from the house, designed to change house air every minute or so. Can be gable mounted with a self-closing louver in an upper hall ceiling.

**Winders**  Pie-shaped treads used when staircases curve or make a turn.

**Window frame**  The framing that surrounds and holds the sash.

**Wood stove**  An insulated freestanding metal wood burning unit with a door.

**Zero clearance fireplace**  A prefabricated insulated metal fireplace unit that can be wall-mounted or freestanding.

# INDEX

**Photo #2**

**Photo #4**

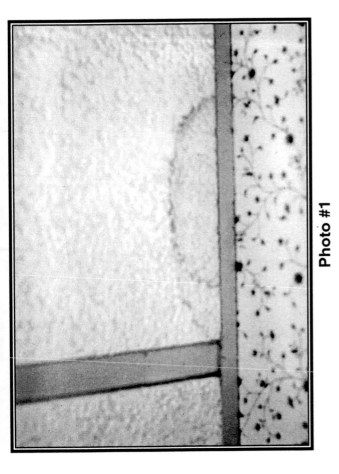

**Photo #1**

**Photo #3**

**Photo #6**

**Photo #8**

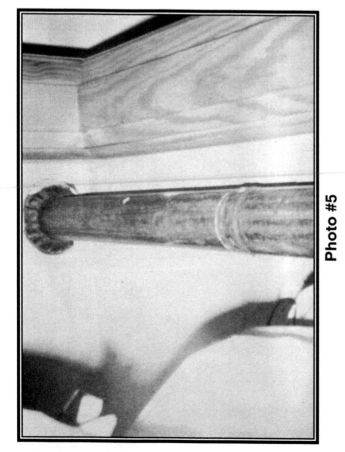

**Photo #5**

**Photo #7**

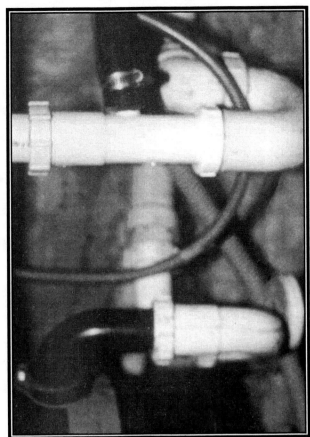

**Photo #10**

**Photo #12**

**Photo #9**

**Photo #11**

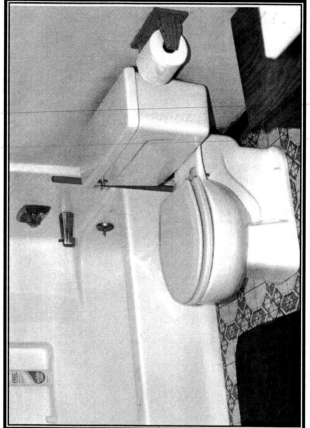

Photo #14

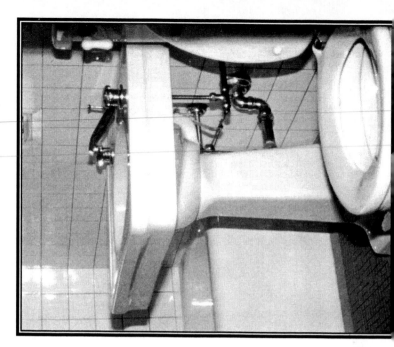

Photo #16

Photo #13

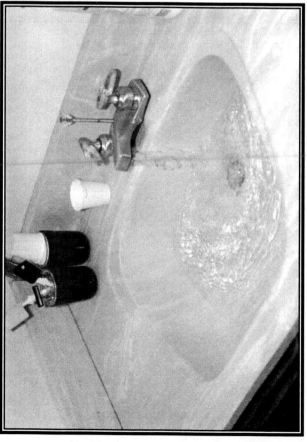

Photo #15

Photo #17

Photo #18

Photo #19

Photo #20

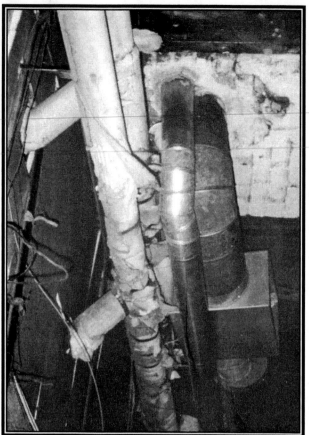

**Photo #21**

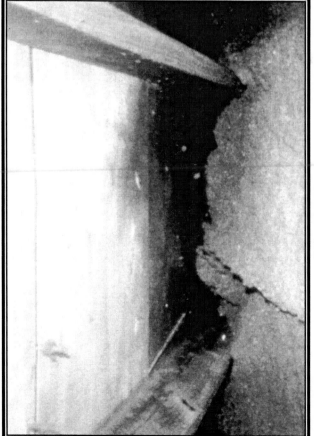

**Photo #22**

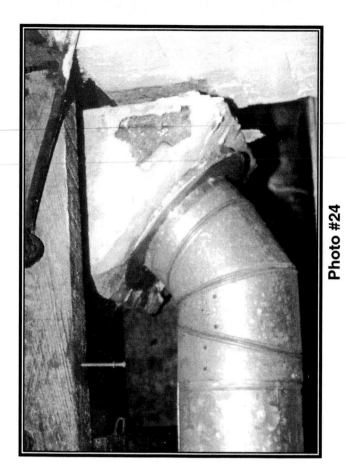

**Photo #24**

**Photo #23**

**Photo #25**

**Photo #26**

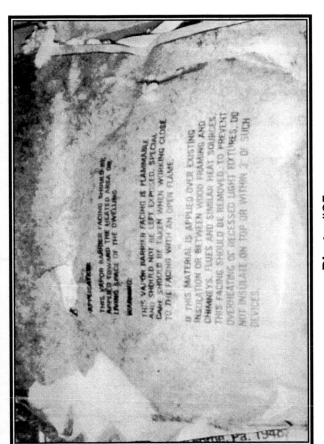

**Photo #27**

**Photo #28**

**Photo #30**

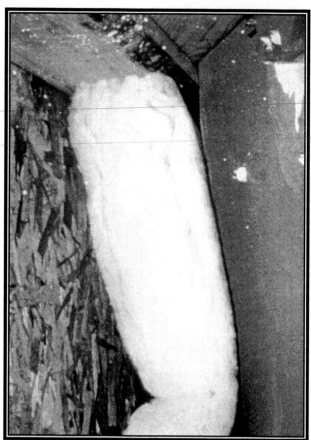

**Photo #32**

**Photo #29**

**Photo #31**

Photo #34

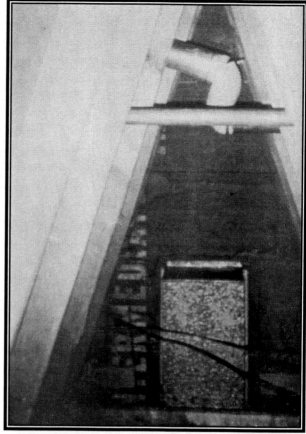

Photo #36

Photo #33

Photo #35

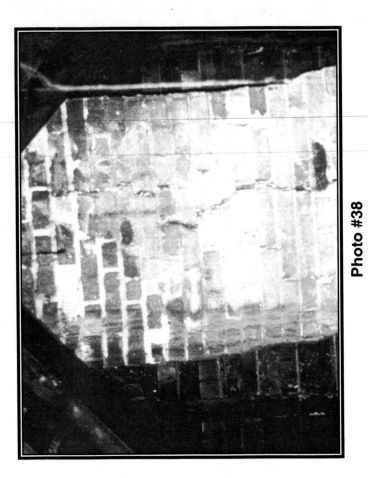

**Photo #38**

**Photo #37**